Children of Other Worlds

'Jeremy Seabrook is one of England's most imaginative and creative writers, with a preacher's talent for prophesy and a capacity for righteous indignation reminiscent of George Orwell.'

Richard Gott, *Guardian*

Other books by Jeremy Seabrook

Notes from Another India

'The book is at its best when describing the collective struggles of the poor [...] there are plenty of insights you're probably not going to find elsewhere.' *Socialist Review*

'This is a powerful book, and justifiably an angry one too – a clear-eyed account presented with sympathy, insight and conviction.' *Resurgence*

Travels in the Skin Trade:
Tourism and the Sex Industry

'A valiant attempt to shine the light of rationality on a subject which has for too long been sensationalised.' *The Lecturer*

'A fascinating, well researched and often passionate book.' *Resurgence*

'Jeremy Seabrook should be applauded for his sensitive presence in the text, avoiding sensationalism or judgement.' *Orbit* (VSO)

Children of Other Worlds
Exploitation in the Global Market

Jeremy Seabrook

Pluto Press

LONDON • STERLING, VIRGINIA

First published 2001 by Pluto Press
345 Archway Road, London N6 5AA
and 22883 Quicksilver Drive, Sterling, VA 20166–2012, USA

www.plutobooks.com

British Library Cataloguing in Publication Data
A catalogue record for this book is available from the British Library

Library of Congress Cataloging in Publication Data
Seabrook, Jeremy, 1939–
   Children of other worlds : exploitation in the global market /
Jeremy Seabrook.
      p. cm.
      ISBN 0–7453–1396–5 (hard) — ISBN 0–7453–1391–4 (pbk.)
   1. Child labor. 2. Child slaves—Employment. 3. Children—Social
conditions. 4. Children's rights. I. Title.
      HD6231 .S4 2001
      331.3'1—dc21
                                                      00–010730

ISBN 0 7453 1396 5 hardback
ISBN 0 7453 1391 4 paperback

10   09   08   07   06   05   04   03   02   01
10   9   8   7   6   5   4   3   2   1

Designed and produced for Pluto Press by
Chase Publishing Services, Fortescue, Sidmouth EX10 9QG
Typeset from disk by Stanford DTP Services, Northampton
Printed in the European Union by TJ International, Padstow, England

Preface

This book is a reflection on children and their social function, drawing largely upon a comparison between industrial Britain in the early nineteenth century and present-day Bangladesh. It is not intended to offer a comprehensive view of child work – in any case, the literature on it is already so extensive that I would hesitate to add to it, if I did not feel that there are other ways of approaching the issue than the pietistic (saving the children), prejudged (children should never work), fatalistic (the children of the poor must work to support their families), bureaucratic (it may be possible to eliminate some of the worst abuses) or economistic (when countries get rich as we have become rich, child labour will wither away).

It is a commonplace that the satanic mills of the early industrial era in Britain have been relocated in the Third World. My starting point was that it might be useful to observe the similarities in the lives shaped by them, both historically and in the present time. The obvious question is what are the effects of another culture, another tradition, another climate? How does child labour in a South Asian, predominantly Muslim, country at the beginning of the twenty-first century differ from child labour in a cold Christian land of the early nineteenth? I have also pursued the similarities between the ravages of free markets in early nineteenth-century Britain and contemporary Bangladesh and the slave plantations of the Caribbean and North America: the imagery of slavery recurs too often within the unfolding drama of globalisation to be ignored.

It is also perhaps worth pausing to wonder whether the removal of these scenes of desolation from our sight has truly liberated us into a 'post-industrial' society; or whether our existence continues to be influenced in one way or another by damaging labour markets that have not ceased to exist, but have merely been removed to another place in the world in the shifting global decor – beneath the appearance of continuous change, an enduring social and economic system persists.

Many people have helped in the making of this book. I would like in particular to thank A. Sivanandan and the staff at the Institute of Race Relations in London, Ian Jack of Granta, Therese Blanchet for her book, *Lost Innocence, Stolen Childhoods* (1996), Duncan

Green of Cafod, especially for his book, *Hidden Lives*, Helen Rahman of SHOISHOB in Dhaka, Tanbir ul Islam Siddiqui of the Underprivileged Children's Education Programme in Dhaka, Rauf Bhuiyan of the NAYAN Foundation, Stuart Rutherford, Nikki van der Gaag and the staff at the *New Internationalist*, my late friend Winin Pereira from Mumbai, Mary Assunta of the Consumers' Association of Penang for her work on the exploitation of children by consumer markets, Subbarow, Uma and Evelyne Hong, P. Rajamoorthy, Anne Beech at Pluto Press for her continuing support, Barry Davis for his constructive criticism, and especially Iqbal Hossein, to whom, with all the other children whose lives are reflected here, this book is dedicated.

Some brief passages from the book have appeared in *Race and Class, Financial Times, New Statesman, Third World Resurgence* and *Third World Network*.

Jeremy Seabrook

Chapter One

Much of the debate about child work and labour has been determined by the West, even though the vast majority of child workers are in the South. This should not surprise us. Globalisation was not chosen by the South, any more than the discourse to which it gives rise. The discussions about children and work are likely to be conceived and expressed from a Western perspective; or perhaps we should say Western perspectives, since there is now an increasingly subtle and nuanced view on children and their rights to protection and self-determination.

It is almost inevitable that the West should demonstrate its expertise in these matters. After all, the arguments were well rehearsed here from the beginning of the industrial era. The defenders of child labour in early nineteenth-century Britain had a ready rationale for its necessity, just as the abolitionists developed a clear justification for why children should be released from the most onerous, dangerous and degrading occupations.

The children of the poor were seen as workers long before the Industrial Revolution. The industrial system merely provided an opportunity for the more systematic employment of children. In 1796, asking the House of Commons to reject Whitbread's Minimum Wage Bill, Pitt said, 'Experience had already shown how much could be done by the industry of children, and the advantages of early employing them in such branches of manufactures as they are capable to execute' (Hammond and Hammond 1947).

The arguments in favour of child labour echoed those which defended the slave trade: the assertion that if England were to shorten the working day for the labour of children others would gain an advantage, paralleled that which insisted that if England were to abandon the slave trade her rivals would seize it. The violent abduction of slaves from coastal areas of West Africa had its counterpart in the transfer of pauper apprentices from the London workhouses to the cotton mills of Lancashire, who 'are sent off by wagon loads at a time [and] are as much lost to their parents as if they were shipped off for the West Indies' (*Life of Sir Samuel Romilly*, 1842 edition). Few of the asperities imposed by the British on their subject peoples abroad had not already been tried and tested on their own poor.

1

Indeed, children were a significant minority among those pressed into slavery. James Walvin, in *Black Ivory*, records, 'Throughout the era of the slave trade, children made up about 34 per cent of the African population. But before 1800 fewer than 20 per cent of slaves carried across the Atlantic were children. There were striking variations in the proportion of children shipped from different slave-trading regions (perhaps 35 per cent of slaves from Sierra Leone were children, for instance.) But as the slave trade developed, the overall percentage of children found on the slave ships rose quite markedly.'

It seems that as the industrialisation of Britain proceeded at home, with its increasing numbers of women and children in mills and mines, so the numbers of children in slavery also grew. The early industrialisation of Britain coincided with the later years of the slave trade; and echoes and correspondences between the condition of slaves in the sugar islands and of workers in the new towns of Lancashire are unmistakable. The growing clamour against slavery which led ultimately to its abolition was aided in no small measure by the beneficent effects of free labour which were becoming clearly visible at home: poverty, lack of livelihood, economic necessity had such a powerful influence upon the willingness of women, children and men to work in mills, mines and factories that it required no great far-sightedness to observe that the outcome of economic pressure upon free labour was not so different from – and far less troublesome than – the slavery of the plantations. Devotion to free labour may have been encoded in the laissez-faire liberal ideology of the late eighteenth century, but its practical consequences were plain for all to see in the new raw settlements around the mills in the industrial north of England. Perhaps this is why so many of the abolitionists of the slave trade failed to extend their sympathy to the workers in the new centres of industry at home. In other words, the very convergence of the working conditions between slaves in the colonies and free labour at home had emboldened the humanitarians to jettison a dependency upon slavery that was already becoming as unnecessary as it appeared distasteful.

Walvin says:

By the nineteenth century, when the centre of the trade was west-central Africa, there was a dramatic increase in the number of boys being shipped ... It was easier to pack more young slaves tightly into the holds. It might also be the case that younger, healthier slaves were better able to withstand the long trek from their interior homelands to the slave ports on the Angolan coast. However we arrange the figures for the nineteenth century slave trade, we find ourselves staring at children. Between 1811 and

1867 more than 41 per cent of all slaves shipped across the Atlantic were children.

When it became obvious that the traffic in human labour, as well as the slavery which was its objective, were doomed, it is perhaps scarcely surprising that the last replenishment should have been of the very young: it became important to breed a new generation of slaves. Previously, the slaves had been recklessly abused, and a large percentage had died on the journey from Africa and from overwork in the plantations. John Newton, a slave trader who subsequently entered holy orders, said it had been found cheaper to work them to their utmost physical capacity and that 'a slave seldom lived more than nine years after importation'. When the supply of fresh slaves dwindled, the cumbersome and expensive business of ensuring their survival further contributed to the appeal of 'free' labour, for whose welfare employers had not the slightest responsibility. Walvin states, 'On Worthy Park estate in Jamaica, of two batches of Africans bought in 1792, more than half were dead within four years. Their owner, alarmed at the erosion of his costly purchases, shifted most of the survivors to a healthier lower location.'

The slave trade was abolished in 1807, although slavery did not cease for another thirty years. (Of course, it has proved more durable than the abolitionists could ever have imagined and it is extremely flourishing today). The trade being threatened with extinction, the traders were driven to use children as an insurance against the time when they would no longer have a fresh supply of slaves from Africa.

No such exhaustion of supply of child labour affected the industrial system at home. The Hammonds make the connection explicit. Speaking of the last quarter of the eighteenth century, they write:

The needs of the London workhouses on the one hand [i.e. over-flowing with 'surplus' orphaned and abandoned children] and those of the factory on the other, created a situation painfully like the situation in the West Indies ... In the workhouses of the large towns there was a quantity of child labour available for employment, that was even more passive and powerless in the hands of a master than the stolen negro, brought from his burning home to the hold of a British slave ship. Of these children it could be said, as it was said of the negroes, that their life at best was a hard one, and that their choice was often the choice between one kind of slavery and another. So the new industry which was to give the English people such immense power in the world borrowed at its origin from the methods of the American settlements ... These London workhouses could be made to serve the purpose of the Lancashire cotton mills as the Guinea coast served that of the West Indian plantations.

The Hammonds called it 'this child serf system' (Hammond and Hammond 1947). In the eighteenth century, child work was generally and uncritically accepted. Daniel Defoe, in his *Tour Through the Island of Great Britain*, remarked:

> We came to Taunton ... One of the chief manufacturers here told us, that they had eleven hundred Looms going for the weaving of Sagathies, Duroys and such kind of Stuffs; he added, That there was not a Child in the Town, or in the Villages round it, of above five Years old, but if it was not neglected by its Parents, and untaught, could earn its own Bread.

Even after the industrial system had long been established, and its effects were clearly visible, there was no dearth of apologists for the employment of children. Andrew Ure, in his *Philosophy of Manufactures*, stated in 1835:

> I have visited many factories, both in Manchester and the surrounding district, during a period of several months, entering the spinning-room unexpectedly, and often alone, at different times of day, and I never saw a single instance of corporal chastisement inflicted on a child; nor, indeed, did I ever see children in ill-humour. They seemed to be always cheerful and alert; taking pleasure in the light play of their muscles, enjoying the mobility natural to their age. The scene of industry, so far from exciting sad emotions in my mind, was always exhilerating (sic) ... The work of these lively elves seemed to resemble a sport, in which habit gave them a pleasing dexterity. (quoted in Thompson 1963)

Arguments in favour of child labour were twofold. First of all, it seemed entirely natural to those who derived benefit from it that the children of the poor should be set to work, and they saw in any diminution of that benefit, ruin, not only of themselves, but of the entire country. The first Act in Britain which limited the hours of labour of apprentices to twelve per day was declared by the mill owners to be 'prejudicial to the cotton trade', as well as 'impracticable'. To take even an hour or two from the working hours for instruction in reading, writing and arithmetic, would 'amount to a surrender of all the profits of the establishment'. Indeed, there is a special familiarity in our feelings about child labour in the Third World: whether they are harvesting the fruits of the earth for our tables or making our garments, the unfree labour of children in the carceral suburbs of Dhaka or Jakarta rouses powerful memories. Reformers had likened the employment of children to slavery on the West Indian plantations: these two memories fuse as we contemplate (or avoid contemplating) the fate of the children of the Third World. Indeed, the contemporary experience of many child workers in Brazil, Bangladesh, India or Nepal represents an uncanny fusion of slavery and early industrialism; coercive and exploitative, although for the most part the bonds that tether them to their work are those of poverty and necessity rather than leg-irons or other more material instruments of physical restraint; although these are also by no means absent.

When it was suggested, in the first Factory Act of 1802, that visitors should be engaged to supervise the implementation of simple reforms proposed for the well-being of children, the masters declared:

> The mills or factories will become a scene either of idleness and disorder, or of open rebellion; or the masters, harassed and tired out by the incessant complaints of their apprentices, and the perpetual interference of the visitors, will be obliged to give up their works; and some of them may become bankrupts, or be obliged *to remove to a foreign country, leaving their apprentices a grievous load upon the Parish where they are employed.*

The same arguments are deployed two centuries later; evidence, if it were needed, that, whatever changes have occurred in the industrial societies of the West, its economic system remains resistant to all but the most superficial reforms and modification.

Nor were apologists for child labour confined to the early industrial period. As late as 1908, Asa Candler, First President of Coca-Cola, speaking in Atlanta at the fourth annual Convention of the National Child Labour Committee to protest at the horrific conditions where women and children worked a sixty-hour week, breathing cotton motes, for 50 cents a day or less, astonished his listeners when he said:

> Child labour, properly conducted, properly surrounded, properly conditioned, is calculated to bring the highest measure of success to any country on the face of the earth. The most beautiful sight that we see is the child at labour. In fact, the younger the boy (sic) began work, the more beautiful, the more useful his life gets to be. (Pendergrast 1993)

On the other hand, the reformers and abolitionists seem to inhabit a different material and moral universe from that in which the optimistic appreciation of child work appeared self-evident. C. Turner Thackrah wrote in 1832:

> The employment of children in *any* labour is wrong. The term of physical growth ought not to be a term of physical exertion. No man of humanity can reflect without distress on the state of thousands of children, many from six to seven years of age, roused from their beds at an early hour, hurried to the mills and kept there, with the interval of only 40 minutes, till a late hour at night; kept, moreover, in an atmosphere impure, not only as the air of a town, not only as defective in ventilation, but as loaded also with noxious dust ... There is scarcely time for meals. The very period of sleep, so necessary for the young, is too often abridged. Many children are sometimes worked even *in* the night. (quoted in Thompson 1963)

The reports of the Children's Employment Commission in 1843 describe conditions in the calico-printing, lace, hosiery, metal, earthenware, glass, paper and tobacco manufactures. The Commission told of children beginning work at three or four years old in their own homes and at five in the manufactories, and of being in regular employment by the age of seven or eight. The hours of work were twelve, in many instances, fifteen, sixteen and eighteen hours consecutively being common, the children generally working as long as adults. In the majority of cases examined by the Commissioners the places of work were 'very defective in drainage, ventilation and the due regulation of temperature', while 'little or no attention' was paid to cleanliness (Second Report of Children's Employment Commission 1843). 'Where deleterious substances were used there was generally no accommodation for washing or for changing clothes. The privies were disgusting, often the same for male and female' (Second Report of Children's Employment Commission 1843:136).

In evidence to the Factories' Inquiries Commission in 1836, Dr Hawkins, speaking of Manchester, said:

> I must confess that all the boys and girls brought before me from the Manchester mills had a depressed appearance, and were very pale. In the expression of their faces lay nothing of the usual mobility, liveliness and cheerfulness of youth. Many of them told me that they felt not the slightest inclination to play out of doors on Saturday and Sunday, but preferred to be quiet at home. (cited in Engels 1952:158–9)

The Children's Employment Commission (1843) report on the metal industries of the Midlands:

The children are described as half-starved and ragged, the half of them are said not to know what it is to have enough to eat, many of them get nothing to eat before the mid-day meal, or even live the whole day upon a pennyworth of bread for a noonday meal – there were actually cases in which children received no food from eight in the morning until seven at night.

The same Commission showed Boards of Guardians in Staffordshire, Lancashire and Yorkshire still getting rid of pauper boys of six, seven and eight, by apprenticing them to colliers, with a guinea thrown in

for clothes. The boys were wholly in the power of the butties, and received not a penny of pay. One boy in Halifax who was beaten by his master and had coals thrown at him, ran away, slept in disused workings, and ate for a long time the candles that he found in the pits that the colliers had left overnight.

E.P. Thompson, in *The Making of the English Working Class*, quotes a minister of religion in an isolated part of North Yorkshire who related the story of a boy whom he had recently interred who had been found (in the mill where he worked) standing asleep with his arms full of wool and had been beaten awake. On that day he had worked for seventeen hours. He was carried home by his father, was unable to eat his supper, awoke at four o'clock the next morning and asked his brothers if they could see the lights of the mill as he was afraid of being late, and then died.

In the Britain of the early nineteenth century, as in the Bangladesh of the late twentieth, it was economic pressure which made parents overcome their reluctance to allow their children to enter the factories. As the income of the Lancashire handloom weavers declined, families could make a living, at first only by employing their own children, and then by sending them into the factories – a practice common all over the subcontinent today. Robert Owen stated that many children were employed under the age of six, some only four or five. In one instance, he had heard of a baby of three working.

Infants in the factories picked up waste cotton from the floor. Being able to crawl under the machines, the smaller they were, the better. Three-quarters of the children in mills were 'piecers', joining threads that had broken in spinning machines. In Dhaka, the children in the garments factories were – and in some cases, continue to be – 'helpers', working at similar tasks; sweeping up waste, cutting threads, sewing buttons as they sit on the floor beneath the Juki or Brother sewing machines at which their older sisters are working. Keeping children awake was a problem, then as now. Use of a stick was necessary in some mills. Contemporary employers in Jakarta and Bangkok sometimes place amphetamines in the drinking water to the same end. The principle is constant; only the technology has become more sophisticated.

In *The Rise of Modern Industry*, the Hammonds tried to capture the quality of how it must have felt to the poor at the time of the Industrial Revolution. They say that whatever the upheavals which affected medieval Europe, or the Europe of Philip II or Louis XIV, there were great numbers of people whose daily lives were little altered by the revolutions of high politics, even though they doubtless continued to suffer varying degrees of misery according to the folly of their rulers.

> This was not the experience of Englishmen (sic) when the economy that governed the life of the village, part peasant, part textile, was merged in the new system of capitalist agriculture and the new system of factory production. A man's life was profoundly altered in its reach, its habits, its outlook, its setting, when, from being some kind of craftsman or a peasant with various tasks and interests, he became a unit in a series of standardized processes. The lives of women were not less intimately affected. In the economy by which the family was provided with food and clothing before the Industrial Revolution, woman's share was definite and visible. Women spun and wove in their homes, brewed the ale, looked after the pigs and fowls; their functions, if different from those of their husbands, were not less important. Specialization extinguished this life, and the women who helped to spin and weave the nation's clothes under the new system, left their homes for the factory, where they found themselves in competition with men, working under disadvantages so easily exploited by their masters that the law treated them as young persons in order to protect them. Large numbers of men and women lost their chief shelter, for in the eighteenth century custom was the shield of the poor, as the law was the weapon of the rich. The poor were thrown into an unfamiliar world where they had neither tradition nor experience to help them. This impression of the age [was] an age of revolution, a migratory society ... The powerlessness of public opinion or settled law is illustrated by two incidents of the time: a hundred men could be killed in a colliery accident in Northumberland without a coroner's inquest; the apprentice children who had been collected in a Lancashire mill could be cast adrift on the sands by their master to beg or steal or starve among strangers without the intervention or notice of magistrate or law. (pp. 242–4)

The Hammonds, like many other defenders of the poor, have been accused by their critics of exaggerating the sufferings of those they described. In fact, their work – and that of other ostensibly partisan historians – is vindicated in some measure by the verification of eerily similar conditions today, in the here-and-now and all over the world. If human beings can be diminished, damaged and disregarded to the degree that is happening in South Asia now, nothing should strike us as implausible in contemporary accounts of how the possessing classes behaved towards the poor in the early industrial era.

The violent wrenching of children from their parents to serve industry parallels the justifications heard for the slave trade. Romilly, describing a speech made in the House of Commons in 1811, said:

> Mr Wortley ... insisted that, although in the higher ranks of society it was true that to cultivate the affections of children for their family was the source of every virtue, yet that it was not so among the lower orders, and that it was a benefit to the children to take them away from their miserable and depraved parents. He said too that it would be highly injurious to the public to put a stop to the binding of so many apprentices to the cotton manufacturers, as it must necessarily raise the price of labour and enhance the price of cotton manufactured goods.

Significantly this form of 'transportation' was reformed in 1816, by which time steam power was taking over from water power and mills could be built closer to the larger centres of population.

The same arguments were deployed by the defenders of the slave trade.

> It was contended, and attempted to be shown by the revival of the old argument of human sacrifices in Africa, that [the slaves] were better off in the islands than in their own country. It was contended, also, that they were people of very inferior capacities, and but little removed from the brute creation; whence an inference was drawn that their treatment, against which so much clamour has arisen, was adapted to their intellect and feelings.

In order to understand the past, the Hammonds had both historical documents and the oral transmission of the experience of industrialisation as well as of the slavery which was its precursor. We have an even more accessible – though rarely visited – source: for the current lives of the migrants to the cities of Asia, Africa and South America are, to a degree difficult to overstate, repeating our story. Cultural differences are great, but the material compulsions of poverty, overwork and want readily overshadow them. An Islamic society, an ancient and rooted Bengali culture, it might be thought, would uniquely inflect the experience of urbanisation and modernisation in Bangladesh. What should not surprise is how the need to survive strips away culturally specific practice and tradition, and plunges humanity into an instantly recognisable and shared predicament. Culture is, at best, a flimsy and threadbare protection against material want.

By listening to these contemporary testimonies, we may recover something of the apprehension, the fear and hope, the driven and tormented upheavals through which our own people passed not so long ago. If we are alert also to the forced migrations of the slave trade, the kidnappings, the abductions and transport of young people to an alien environment in which they were forced to labour for the profit of others, this can only sharpen our understanding, not only of our own past, but also of what life is like for many young people in the Bangladesh of the turn of the century.

For migrants are now characteristic figures of our time, and the epic change that transformed Britain two hundred years ago has its parallel in present-day countries of the South. Bangladesh is also – as was Britain in its time – a primary exporter of labour to South and East Asia and the Gulf. Bangladeshis are to be seen on the rubber and palm-oil plantations of Malaysia, occupying the low-paid jobs in the squalid barracks lately occupied by Tamils taken there four or five generations ago by the British. Bangladeshis live in the cockroach-infested kitchens of restaurants all over Europe and North America, they have made their home in the elegant decay of Spitalfields and the squalor of Brick Lane, they lodge on construction sites in Thailand, Laos and Cambodia; they work as drivers in Saudi Arabia, as cooks in Dubai, as domestic servants in Kuwait, Abu Dhabi and Dubai. In the process, they are regularly cheated, swindled and robbed wherever they go.

One of the most pathetic spectacles that I saw on my travels for this book was a group of Bangladeshi 'overstayers' manacled together like convicts at Kuwait airport. They were being marched under armed guard to the flight back to Dhaka. As they passed through the marble corridors, Kuwaitis stopped to look and children chanted 'Bangla Bangla' as though this were a term of abuse. Once on the plane, they were released from their handcuffs. They told a different story. Paid several months in arrears, they had been advised by their employers that there was no need to renew their visa, since they (the masters) would take care of it. The employers then denounced them to the authorities who picked them up and deported them in their daily swoops on 'illegals'. The employers gained the benefit of several months' free labour.

Dhaka, November 1999. It was a day of *hartal*, or political strike, called by the Opposition Bangladeshi National Party, which closes down the country from dawn until dusk. No cars or buses on the streets, only the rattle of the cycle-rickshaws and the ringing of their bells, which lend the city a strangely pre-industrial air. Groups of young men – some of them little more than children – hired by political activists patrol the streets, chanting, menacing. If they meet any motorised vehicle, they turn out the occupants, overturn it and set it on fire. Businesses are shuttered; only shops selling food and medicine are allowed to open, and many of these trade fearfully through half-closed grilles.

I sat in the lobby of the hotel, where I met a contractor who had come to recruit labour for his garment factory in the United Arab Emirates. He said:

I place an advertisement in the paper and five thousand people will come. I take 200 of them, fly them to Dubai. They live in barracks and work in the factory. Twelve, fifteen hours a day. They sleep and eat. Maybe they dream, it's up to them. They are in a foreign country, the language of which they do not know, the customs of which are unfamiliar to them. I pay them enough to send home to their families. They are captive. This is called slavery, but it is legal.

Within Bangladesh itself, migration has extended far beyond the annual procession of people displaced by the natural fluidity of land and water: more than 1.5 million young women and children entered the garments industry in the 1980s and 90s. In 1983 there were fifty garment factories. By 1997, 2,400. The number of workers had risen from 10,000 in 1983 to 1.3 million by 1997. In the early morning Dhaka becomes a city of women as they emerge from the slums in vivid colours of violet and lime green, orange, crimson and gold, a music of bangles and a kicking of dust from their chappals on the margin of the road, an extraordinary frieze of youth swallowed up by the factories with their white strip lighting, overcrowding and excessive hours of labour. They reappear twelve or fourteen hours later, the brightness of their clothes dimmed by the sodium street lights, their eyes ringed with fatigue, walking in groups so that they will not be attacked on their journey back to the slums. These are, overwhelmingly, the daughters of farm workers, children of the rice fields, part of the great movement of humanity which has always taken place in the country as the rivers move and eat up the land in the delta, but which has become a vast transfer of people to the city with the opening up of factory work in Dhaka and Chittagong. The population of Dhaka has almost doubled in twenty years – a faster rate of growth than anything experienced by England in the early nineteenth century.

The comparison of the child workers in Britain to slavery in the sugar islands was not simply a metaphor; and neither is it only a picturesque description of the widespread abduction of children in South Asia for the purposes of prostitution and other forms of degrading labour. Shima das Shimu works with UBINIG, a non-governmental organisation in Dhaka, on the trafficking of children.

> Many girls come to Dhaka and they get lost. They may get trafficked for the sex trade or the trade in organs, or they are abducted for marriage. Sometimes whole families – that is, women and children – are trafficked. They are taken to the border, and once they cross, the trafficker disappears and the victims go to various destinations. Children between the ages of nine and sixteen are very valuable. The price varies – from 10,000 taka to 5 lakhs (US$ 400–10,000). The network trafficking children from Jessore to India has been traced, but it is very difficult to stop free movement across the border because by doing so you impede people going for family reasons, for legitimate work.

Shirin and Ranwu live in a small town not far from Khulna, close to the Indian border. Their neighbour, a graduate who gave 'tuitions' to children to help them in school, suggested to their parents that he take them to Dhaka for the wedding of a member of his family. They would be paid a small sum for their help with the arrangements for the wedding party. Shirin was then thirteen and Ranwu twelve. They did indeed go to Dhaka, but were taken back by minibus by a different road to the Indian border. The parents, alarmed after a silence of two weeks, alerted the authorities. By this time, the girls had been sold. They had been taken to Calcutta and then by train to Mumbai.

Their father, Maksud, is a rickshaw driver. The family owns a little land. He had never been further than Khulna in his life, but he resolved to find his daughters, even though at that time he had no idea of their whereabouts. His fifteen-year-old son went with him. He sold the land and located the red light district in Calcutta; when showing photographs of his daughters, he met a woman who informed him that she knew who had taken them to Mumbai. This contact said she knew they had been sold to a brothel in Mumbai, but denied any part in the transaction. The father and son took up their position day and night on the streets behind Mumbai Central station. One day, the brother thought he saw Shirin in a *tonga*, a horse-drawn vehicle. They jumped into a taxi, but were held at the traffic lights and the *tonga* disappeared. This gave them the courage to keep on searching.

The buildings around Foras Road, in the district known as Kamathipura, are densely built, with hidden rooms, secret passages, locked roof terraces, so it is easy to hide kidnapped children. A number of police raids helped release other girls, some of them maintained in the most abject conditions as a way of disciplining them. Many of the girls are illiterate – once they have been separated from their families, they have no means of communicating with them. It later emerged that the two girls were being kept until they were a little more mature, so that their virginity would command a higher price. It is no use presenting girls who are crying and weeping as they will not please potential customers. They have to learn the arts of pleasing. Once employed, they quickly lose value.

One night, the father thought he caught a glimpse of his daughter behind the barred windows of the little blue- and green-washed rooms, with their white strip lighting. Shirin had been sleeping. Something told her to go to the window. She cried out to her father, who was standing on the opposite side of the narrow street. Pretending to be a client, he was shown into the room where eight or nine girls were kept. The girls had not been raped and they still believed they were being trained to be servants in the house of a rich family. The police came and beat up the brothel-keeper, but there was no attempt to investigate the kidnappers. The brothel-keepers are only the end of a long chain of dealers. Shirin and Ranwu returned home. The teacher has never reappeared. His mother is still a neighbour of the family whose girls were stolen. Nothing is ever said about what happened. Both girls are now married.

It is the credulity of country people today, their unfamiliarity with the wiles of the powerful which makes it easy to entrap children. Promises of work – almost any work – are sufficient to induce them to part with their daughters. They are rarely kidnapped by force, as slaves were at an earlier time, even though their fate may scarcely be better. The story of Shirin and Ranwu is unusual in that the outcome was happier than might have been expected. Richard Hart gives this insight into the state of mind of the stolen child.

A revealing picture of the impact of the slave trade in the mid-eighteenth century on a remote inland Ibo community is contained in the memoirs of the ex-slave Olaudah Equiano. Born in 1745, he was kidnapped by African slave traders at the age of eleven, and sold and resold several times until he reached the coast seven months after his capture. Terrified of the white men

who came to inspect him and convinced that, unlike his own
people, they were cannibals, he was taken aboard ship and resold
in Barbados. He later worked as a slave in Virginia and was
resold to a naval officer and worked aboard ship, and was subsequently resold to a Philadelphia merchant who owned
schooners trading to the Leeward Islands. By doing a little
trading on the side Equiano raised the forty pounds needed to
buy his own freedom in 1766. Thereafter he worked on
merchant ships until 1777, finally settling in England where he
became active in the abolitionist cause.

It is significant that the word 'abolitionist' was used by those
working against indentured child labour in Britain, as by those
committed to ending both the slave trade and later, slavery itself;
and it is still the term of preference among those seeking to
eliminate child labour today. There is in this perhaps a recognition that these evils all have a common source, namely the
dehumanising of human beings, whether by race or by age. We
might also add 'by class'. Eric Williams, in *Slavery and Capitalism*,
notes that

> The first slaves [in the Americas] were Indian. They succumbed
> to the excessive labour demanded of them, insufficient diet, the
> white man's diseases and their ability to adjust themselves to the
> new way of life ... The immediate successor of the Indian was
> not the Negro but the poor white. White servants – some
> indentured, signing a contract before leaving that they would
> serve for a stipulated time in return for their passage. Others
> were convicts to serve for a specific period ... In 1683 white
> servants represented one-sixth of the population of Virginia.
> Two-thirds of immigrants to Pennsylvania during the eighteenth
> century were white servants. Kidnapping became a regular
> business in London and Bristol. Adults would be plied with
> liquor, children enticed with sweetmeats. Convicts provided a
> steady stream of white labour. The harsh feudal laws of England
> recognized three hundred capital crimes ... It has been suggested
> that it was humanity for his fellow-countrymen and men of his
> own colour which dictated the planter's preference for the Negro
> slave. Of this humanity there is not a trace in the records of the
> time, at least as far as the plantation colonies and commercial
> production were concerned ... In Maryland servitude developed
> into an institution approaching in some respects chattel slavery
> ... On the sugar plantations of Barbados the servants spent their
> time grinding at the mills and attending the furnaces, or digging
> in this scorching island, having nothing to feed on (notwith-
> standing their hard labour) but potato roots, nor to drink, but

water with such roots washed in it, besides the bread and tears of their own afflictions; being bought and sold still from one planter to another, or attached as horses and beasts for the debts of their masters, being whipt at the whipping-posts (as rogues), for their masters' pleasure, and sleeping in sties worse than hogs in England.

In Britain, the system of pauper apprentices in factories declined after legislation against it in 1816, but children continued to be employed as wage-earners, and the conditions of their working life differed little from the earlier period – much as slaves continued to work on plantations once the trade had been (officially) banned. The following passage describes the experience of a ten-year-old parish apprentice at Litton Mill in Derbyshire in 1815–16:

The Prentice House was a large stone house surrounded by a wall from two to three yards high with but one door, which was kept locked. It was capable of lodging about one hundred and fifty apprentices.

We went to the mill at five o'clock without breakfast, and we worked till about eight or nine, when they brought us our breakfast, which consisted of water-porridge with oat-cake in it and onions to savour it, in a tin can. This we ate as best we could, the wheel never stopping. We worked on till dinner time, which was not regular, sometimes half-past twelve, sometimes one. Our dinner was thus served to us. Across the doorway of the room was a cross-bar like a police bar, and on the inside of the bar stood an old man with a stick to guard the provisions. These consisted of Derbyshire oat-cakes cut into four pieces, and ranged in two stacks ... the one was buttered and the other treacled. By the side of the oat-cake were cans of milk piled up – buttermilk and sweet milk. As we came up to the bar one by one the old man called out, 'Which'll ta have, butter or treacle, sweet or sour?' We then made our choice, drank down the milk and ran back to the mill with the oat-cake in our hand without ever sitting down. We then worked on till nine or ten at night without bite or sup. (Harrison 1984)

Britain was not alone in the mistreatment of its child workers in the early industrial era. Fielden, in his *Curse of the Factory System*, quotes a pamphlet published in 1833, in which an imaginary visitor is being shown over the American mills:

He might see in some, and not infrequent, instances, the child, and the female child, too, driven up to the clockwork with the cowhide, or the well-seasoned strap of American manufacture.

We could show him *many* females who have had corporeal
punishment inflicted upon them; one girl, eleven years of age,
who had her leg broken with a billet of wood; another who had
a board split over her head by a heartless monster in the shape
of an overseer of a cotton-mill... (cited in Hammond and
Hammond 1947:209)

Arif was ten years old when he went to live in a garment factory in
Old Dhaka with six other children. His sister was working a
machine, while he and the other children worked as helpers. They
were allowed to sleep in the factory at night as a privilege, but they
also served as its guardians, unofficial and unarmed nightwatch-
men. Most of the children preferred staying in the factory to going
home, since there were bundles of cloth to sleep on and they didn't
have to travel to or from work. The factory owner lived nearby and
took care of their needs. Arif said:

His servant used to bring us *roti* and tea in the morning at six
o'clock. She was only a young girl and she didn't like coming
into the factory because we were all boys there. She used to leave
the tray downstairs and we went to take it. Most days my sister
would bring some tiffin from home, but we were also given
vegetables and rice at midday by the owner. It was always
potatoes or *baigan*, all heaped together on one plate, so we often
used to fight over the food. I often didn't eat it because I had
some from home. After that, there was nothing else except a *roti*
sometimes. If we had any money we bought a packet of biscuits.
I often fell asleep working. We were beaten by the overseers or
the other workers if we didn't keep up with the work. We sat
under the machines, cutting threads, tying knots, sewing
buttonholes and buttons. Sometimes we did not go outside for
a week. When the time came for children to leave the factories,
we were told one morning we had to leave that same night, even
if we had nowhere to go. The *malik* gave us 200 taka each (about
US$ 4). I had one friend who had no family in Dhaka. He came
to my house in Mirpur, but there was no room for him to stay.
Next day he went back to his village.

Here is an account by James Walvin in *Black Ivory*:

Occasionally we catch a passing glimpse of the otherwise
anonymous slaves. Clarissa was among the batch of Africans
brought to Worthy Park [in Jamaica] in 1792. Described as
'Congolese', it is hard to prove her exact origins, but she was
very young, perhaps only thirteen when captured and transported. Clarissa, like her shipmates, was turned loose on various

tasks on the estate, and in its less demanding provision grounds. When she regained her strength, she was put into Worthy Park's Second Gang, acquiring a new name, Prattle. A mere three and three-quarters years after arriving in Jamaica, she succumbed to the flu and died at the age of seventeen.

Factory Acts designed to improve the conditions under which children worked were passed early in the nineteenth century (1802, 1819, 1831), but these were generally ignored. The report on Wigan of two Lancashire Visitors of Cotton Factories appointed by the Quarter Sessions in 1828, stated: 'The children go to work at five in the morning and continue till nine at night. Some few are allowed to go to their homes to breakfast and dinner, but by far the greatest number are not suffered to go out of the premises at all between the hours mentioned.' The Visitors pointed out that the Act could not be enforced, since the Wigan magistrates were all cotton mill proprietors and were forbidden to try cases under the Act (Hammond and Hammond 1947:202). Only after the report of Michael Sadler's parliamentary committee in 1832 did the Factory Act of 1834 prohibit the employment of children under nine (except in silk mills) and limit the work of children between nine and thirteen to forty-eight hours per week, with a maximum of nine hours in any one day, and that of those between fourteen and eighteen to sixty-nine hours a week, with a daily maximum of twelve hours. This, too, was widely evaded. The Ten-Hour Bill was passed only in 1847.

In the same year that the Factory Act of 1834 was passed the Poor Law Amendment Act also became law. The words of the Commissioners reveal the social prejudice of authority more than they illuminate the failings of the previous system which they found to be

A check upon industry, a reward for improvident marriage, a stimulus to increased population, and a means of counterbalancing the effect of an increased population upon wages; a national provision for discouraging the honest and industrious, and protecting the lazy, vicious and improvident, calculated to destroy the bonds of family life, hinder systematically the accumulation of capital, scatter that which is already accumulated, and ruin the taxpayers.

The response of the Commissioners echoes that of the colonial planters. Ten years earlier, in 1823, the British government had adopted a new policy of reform towards West Indian slavery. The policy was to be enforced by orders in council, in the Crown Colonies of Trinidad and British Guyana; its success, it was hoped, would encourage the self-governing colonies to emulate it spontaneously. The reforms included: abolition of the whip; abolition of the Negro Sunday market by giving the slaves another day off to allow time for religious instruction; prohibition of the flogging of female slaves; compulsory manumission of field and domestic slaves; freedom of slaves born after 1832; admissibility of evidence of slaves in courts of law; establishment of savings banks for slaves; and the appointment of a Protector of Slaves whose duty it was, among other things, to keep an official record of the punishments inflicted on the slaves.

The reply of the planters, in the Crown Colonies as well as in the self-governing islands, was an emphatic refusal to pass what they considered 'a mere catalogue of indulgences to the Blacks. Not one single recommendation received the unanimous approval of the West Indian planters ... From the planters' standpoint, it was necessary to punish women. Even in civilized societies, they argued, women were flogged in the Houses of Correction in England. "Our black ladies," said Mr Hamden in the Barbados legislature, "have rather a tendency to the Amazonian cast of character; and I believe their husbands would be very sorry to hear that they were placed beyond the reach of chastisement." A Trinidad planter called it "a most unjust and oppressive (sic) invasion of property" to insist on a nine-hour day for full-grown slaves in the West Indies, while the English factory owner could exact twelve hours' labour from children in a heated and sickly atmosphere...' (Williams 1945).

Today, official responses to suggestions for the reform of social evils are more subtle, although popular indifference may be broadly similar. Confronted by child labour in Bangladesh, the Bangladeshi government readily signs up to all international agreements, protocols and conventions on the protection of children; and these, in general, fail to be observed. International agencies now react to this apparent indifference with a new *flexibility*, a realism, a recognition that the situation on the ground is not amenable to abolition, but that there must be a gradualist approach, an attack only upon the worst kinds of child labour. By a curious historical twist, the exploitation of children becomes the inescapable *economic* necessity for child labour; and like all other ills, the rising prosperity and development of the market economy is seen as the panacea that will eliminate all these unfortunate archaisms, these survivals from another age. In other words, the remedy for exploited childhood is written into a predetermined future, as Bangladesh, India, Brazil and Indonesia become *more like us*. In other words, the answer lies in the benign and inevitable furtherance of globalisation.

Globalisation is a euphemism; but the concept was as familiar in the age of an imperialism which trafficked slaves from Africa to North America and the Caribbean, which took cotton from the slave states (in 1780 the West Indies supplied Britain with two-thirds of its raw cotton) and from India to Manchester, and sent out goods all over the world, even to India, where the far superior indigenous industry had been rigorously suppressed. The British imperial adventure represented a series of efforts to link large parts of the world together in a common destiny – early and violent experiments in globalisation. It is an epic paradox that the British government was compelled to nationalise the East India Company, the first truly transnational company, in order to curb its excesses. Today, it is reckoned the highest policy of the global institutions that governments all over the world should set free similar private entities to work their mischief at will; and there is not, as yet, much sign that any other authorities have the power to mitigate it, even though there is increasing popular revulsion against the growing concentrations of transnational corporations, which control more wealth than nation states.

Since globalisation means neither more nor less than that every country in the world should follow the Western path of development, it is only natural that the arguments deployed in nineteenth-century Britain for and against child labour should be heard once more, this time on the global stage. Much of the argument of those who wish to remove children from factories is expressed in terms that would have been familiar to nineteenth-century Britain; it reflects the certainty that 'reform' and 'improvement' can follow the same (retrospectively) fairly simple path pursued by the first great industrial power. Hayek claimed that the parliamentary investigations of the 1830s and 1840s were evidence of the growing humanitarianism that comes with increasing wealth. E.P. Thompson, in his *Making of the English Working Class*, contests this, asserting that such official inquiries were often only a means of delaying action on known evils. Faith in the benign long-term effects of wealth creation is all very well, but without watchful and unceasing struggle it is unlikely to benefit greatly those out of whose misery it is made.

In any case, although conditions in present-day South Asia may resemble those in early Victorian Britain, the wealth that will both make conscience more tender and provide practical opportunities for the betterment of the people is scarcely available. These discussions are haunted by a sad paradox: how far did the extractive project of Empire contribute to the rise in well-being of the poor in Britain? And if this was indeed a significant element in the improvements enacted here, where will countries like Bangladesh, India or Indonesia find the wealth that will enable them to keep their rich in the state to which they have so swiftly become accustomed and, at the same time, raise up the standards of their poorest? These questions hover over all the assumptions made about economic growth in perpetuity, the model that institution-alises on the global scale the proposition that only if the rich are permitted to become infinitely richer will the poor become a little less poor; and then only relatively. If the conditions in which the poor must labour – especially the most vulnerable, the children of the poor – come to resemble the direst slavery, this is, perhaps, a direct consequence of the effort to follow alien models without the vast colonial hinterland which the British had at their disposal for two hundred years.

And indeed there is evidence, from all over the world, that the wealth of 'independent' and 'sovereign' countries may be increased only by means of intensifying pressure upon the resource base and upon the people who have traditionally depended upon it – the Indians of Brazil, the indigenous minorities of Indonesia, Malaysia, the Philippines; the Adivasis of India and Bangladesh. A caricature of colonialism, a concentrated brutality towards those who stand in the way of the expansion of countries compelled to compete in the global market. In vain, the native peoples of the world protest that they are the representatives of ancient sustainable economic practice. Their way of life, instead of being looked to as an inspi-ration for less damaging ways of living on the earth, is everywhere being wiped out for the sake of immediate short-term gain.

Chapter Two

The evolving discussion in the South has been profoundly influenced not only by the rhetoric of the West, but equally by its actions. One of the most significant moments in the last decade was the threat of the Harkin Bill in the US Congress, which would have banned the import into the USA of all garments produced by child labour in Bangladesh. Indeed, the debate about 'core labour standards' proposed by the United States government at the time of the disastrous World Trade Organisation meeting in Seattle late in 1999 was one of the principal reasons for the chaos which engulfed the negotiations.

Since Bangladesh was at the time of the Harkin Bill the seventh largest exporter of garments to the US, this had the effect of emptying the factories of child workers very quickly indeed. The subsequent package of education and monetary compensation to the families who would be disadvantaged by the loss of earnings of their children was too little, too late, since by the time this became available, as many as 75 per cent of children working in the factories had already been dismissed. Although no research was actually carried out into the fate of these children, it was widely asserted – perhaps exaggeratedly – that many had gone into worse occupations than factory labour. Jehanara Begum, trade union organiser with the Garments Union, asserted that 'at least in the factory adults could keep an eye on the well-being of the children, could make sure they were not over-exploited or abused. When those children went onto the streets to sell flowers and trinkets, into private domestic service, or into factories not making objects for export, who then could know what happened to them?' In her words are echoes of the observations of E.P. Thompson, who admitted that 'while a few of the operatives were brutal even to their own children, the evidence suggests that the factory community expected certain standards of humanity to be observed'. He asserts that, far from the workers behaving unkindly to each other, 'it was the discipline of the machinery itself, lavishly supplemented by the driving of overlookers ... which was the source of the cruelty' (Thompson 1963:373–4).

Certainly, according to Helen Rahman, founder of SHOISHOB, a non-governmental organisation working with domestic servants, the age of domestic servants fell with the exodus of children from the factories of Dhaka. And it is a matter of record that domestic labour is, for many girls between the age of eight and thirteen, a not particularly tender experience. As many as one-quarter in a survey reviewed by Therese Blanchet had been abused at work (Blanchet 1996).

It soon became clear that the foreign abolitionists behind the Harkin Bill – whether their motives were truly humanitarian or whether this had been a cover for US protectionism – had acted over-hastily. The idea that children should be 'rescued' from factory labour, even if the arrangements for their welfare had been put in place before they were dismissed, could affect only a small number of working children, since the great majority are employed in the informal sector, particularly in agriculture, as well as (for girls) in their own homes and families, in small workshops and manufacturing units, in transport, in selling on the pavements anything from flowers, pens, fruit, matches and vegetables to hand-kerchiefs, toys and clothing.

In *The Journal of the Centre for Social Studies*, Dhaka, in July 1996, Md. Abdur Razzaque and Md. Anisur Rahman, writing of the threat of the Harkin Bill, state:

> The Bill also technically ignored to take into consideration the condition of the working children in the non-traded sector. In this paper we have found that the working environment in the non-traded sector is not better, even worse, compared to the garments sector ... There will be a sudden exodus of child labourers from the traded sector into the non-traded sector. This will surely weaken the already poor bargaining power of the child workers which will, in turn, increase the exploitation of the child workers in these sectors, in terms of lower wages, lengthy hours of work, job insecurity and worsening working conditions. (Razzaque and Rahman 1996:91)

There are, in any case, ways of evading restrictions on the employment of children, in Bangladesh as in early industrial Britain. Many children in Bangladesh remained in the factories if they *appeared* to be older than fourteen. In *Capital*, Marx quoted an English factory inspector who stated:

> My attention was drawn to an advertisement in the local paper of one of the most important manufacturing towns of my district, of which the following is a copy: 'Wanted, 12 to 20 young persons, not younger than what can pass for 13 years. Wages 4 shillings a week. Apply etc.' The phrase 'what can pass for 13 years' has reference to the fact that by the Factory Act (of 1844),

children under 13 may work only six hours. A surgeon officially appointed must certify their age. The manufacturer, therefore, asks for children who look as if they were already 13 years old. The decrease, often by leaps and bounds, in the number of children under 13 employed in factories, a decrease that is shown in an astonishing manner by the English statistics of the last 20 years, was for the most part, according to the factory inspectors themselves, the work of the certifying surgeons, who overstated the age of the children...

A similar process, made easier by the absence of birth certificates among the workers, was at work in Bangladesh. This allowed for dismissals to take place on the basis of the apparent age of children; a subjective and highly unreliable index of their true years.

Therese Blanchet notes the efforts to remove children from other factory work in the early 1990s. In the *bidi* (local cigarette) factories of Kustia, children under 4 foot 6 inches (in theory, those below the age of twelve) were dismissed. 'Some small children who were orphans and had no other means of livelihood were allowed to remain. After the dismissal, many returned, either through bribery, bringing lunch for their older siblings and staying on, even scaling the factory wall to get in.' Blanchet adds that those working illegally were paid less than those legitimately employed; they therefore subsidised their own illegality (Blanchet 1996). The image of children bribing their way into exploitative labour is an arresting one: it suggests that their perception of their own interests is often at odds with that of those who wish to better them.

The abolitionists in Bangladesh took their cue from nineteenth-century Britain where large numbers of children had been employed in mills, mines and manufactories; and where increasingly interventionist governments were compelled to respond to the promptings of reformers and campaigners against abusive work and in favour of an extension of educational provision for children. In Britain, as wealth increased in the nineteenth century, the removal of children from damaging labour became easier, despite the protests of employers advantaged by the use of nimble fingers and pliant personalities. Even so, children continued to work in the declining agricultural sector and still worked as vendors and hawkers, as well as pickpockets and hustlers, on the streets of London and other big cities until well into the twentieth century. As late as 1913, when Seebohm Rowntree and May Kendall published their study, *How the Labourer Lives*, they found that in a family with two of six children working in rural south-east England, 'There is a deficiency of 24 per cent of protein in the family's dietary and of 17 per cent in energy value ... Children earn a trifle by picking up acorns by the bushel to feed pigs; a boy goes to the rectory before and after school to clean knives, boots and the like ... .' From two daughters away in domestic service 'the mother receives nothing'. They state:

> There is no doubt that the worker in the country feels that he and his are steadily losing ground. It is not only a question of employment here and now; it is a question of the future of the family. Ask any village mother who is ambitious for her boys where she wants them to spend their lives. The answer is generally 'Not on the land – there's nothing for them'. The girls, too, leave home ... Innumerable girls go into service in towns when they are too young and inexperienced to look after themselves and most need home care ... The parents must at all risks dispense with the burden upon their resources of an inmate who is old enough to earn and yet is earning nothing. 'There isn't a girl in the village besides me,' said one bright young woman of eighteen, who was keeping house for her brother and grandfather, her mother being dead. 'They've all gone away somewhere – into service mostly.'

All this presages the common experience of South Asia today. Rural populations have everywhere been under pressure, the costs of growing staple crops have risen, increasing inputs of pesticide and fertiliser are required to maintain the yield of hybrid varieties of rice. Many small farmers cannot keep pace, and more and more members of the family migrate to the towns. In any case, small farmers are inadequately rewarded for the surplus they make since all must bring their crop to market at the same time, thereby depressing the price. It is the middlemen, who buy up the crops and sell them at rising prices in the lean season, who make significant profit. The story of impoverished farmers sending their children, or moving with their families, to the city is achingly familiar – the sense of inescapable duty to parents in sending home small remittances, the irreversibility of the journey, the unfamiliarity with the rituals of industrial life – all this is the same.

The fact that communications are easier, that people can come and go more readily between home and city, conceals the fact that such a journey is more than a simple bus ride from the village to the city: the sensibility, the psyche and the perceptions have all been changed in the shift from agricultural to industrial labour. For most, there is no going back. Although many girls who have been in domestic service or garment factories in Bangladesh are expected to return for marriage, the large numbers who remain in the city and marry cycle-rickshaw workers suggest that their own sense of themselves, their identity and purpose, and the meaning they derive from their work, have all been radically altered by migration. In any case, even those who do not move from the village undergo a drastic re-learning process. Although they remain essentially peasants, they are working in a capitalised agricultural system which creates hybrid people, neither peasant nor industrial, to go with the hybrid crops they harvest.

It is difficult not to be struck by the continuous interest exhibited by the government of Britain in the lives of the poor. Doubtless, some of this came from a persistent anxiety about the true nature of the class of labour which industrialisation had created, and whether this refractory, sometimes radical and often disaffected population, could be reconciled to the society so conspicuously advantaged by the arduous work they were called upon to perform. But the fact remains that there were constant commissions, inquiries and investigations into the way the poor – and their children – lived. Many of these resulted in legislation designed to limit some of the worst abuses, which they did with varying degrees of efficacy; but there is no doubt that they also furnished the ruling class with information on the state of feeling, the temper of those called into existence by the industrial system – for they represented a new sensibility, a quite different form of human being from the declining peasantry and country-dwellers, who were the raw material for what was to become the industrial workers. Whatever the cause – whether altruism or anxiety – a constant stream of detailed knowledge *about* the poor was made available to those who governed them.

No such compulsions are at work on the governments of the South. In any case, worries about their docility and their capacity for disturbing settled patterns of wealth and power have been con-siderably allayed, the more so since the death of the Soviet Union. If a country like Britain, with the gross inequalities thrown up by industrialisation, could accommodate those who once seemed per-manently estranged from the society that had produced them, the poor of India or Thailand, of Indonesia or Bangladesh appear to present no serious threat to social peace, no matter what degree of social injustice prevails. What is more, these countries have been influenced by other social and cultural factors. Ancient inequali-ties and injustices persist, a heritage of caste, destiny and acceptance, which have always relieved their rulers of that concern for their peoples which was at work in the age of improvement in nineteenth-century Britain.

In Bangladesh, feudal mentalities are often carried from village to city: workers and servants are still regarded as an inferior order of human being. 'Do not give them more than five taka,' I was warned by a corpulent businessman being driven by a skinny sweating rickshaw driver who looked as if he was already in the last stages of consumption, 'it spoils the market' – a demonstration, if ever there was one, of the priority of markets over needy human beings. Child servants are expected to sleep on a mat in a corridor or under a verandah; factory workers remain grateful for work that has been 'given' at the most miserable rates of pay and for excessive hours of labour. The treatment of the poor by the middle class in Bangladesh – as though they existed solely as adjuncts to the comfort and well-being of their social superiors – perpetuates relationships that were familiar in Britain well into the twentieth century.

A recurring theme, not only in the industrial history of Britain and contemporary Bangladesh, but even more emphatically in the history of slavery in North America, is the importance of *discipline;* the need to instruct rural people in the rigours of time-keeping, efficiency and subordination. In a way, the eighteenth-century plantation in Jamaica, Virginia or South Carolina was itself a prototype of industrial production. The fact that it dealt with agricultural produce is an ironic detail, given the systematic violence with which slaves – and, later, workers – had to be schooled to their tasks. In any case, now that agriculture has been industrialised globally, the same measures of efficiency, productivity and output are deployed in agriculture.

Slaves, like industrial workers, were infantilised by the harsh tutelage they had to undergo before they could accept their new role in the world. The employment of children made this easier since they had not developed the sometimes mutinous and rebellious habits of adults forced to take up alien labour in a hostile environment. Equally significant is that the only precedent for such impositions was military: the unquestioning obedience, the rigid timetable, the suppression of all individual characteristics. Indeed, the history of both slavery and industrialisation are penetrated by the ghost of a militaristic coercion.

A Scottish clergyman, James Ramsey, who spent twenty years in the West Indies, gave this account of work and discipline in a book published in 1784:

> The discipline of a sugar plantation is as exact as that of a regiment; at four o'clock in the morning the plantation bell rings to call the slaves to the field ... About nine o'clock, they have half an hour for breakfast, which they take in the field. Again they fall to work, and according to the custom of the plantation, continue until eleven o'clock or noon; the bell then rings, and the slaves are dispersed in the neighbourhood, to pick up about the fences, in the mountains, and fallow or waste ground, natural grass and weeds for the horses and cattle. The time allotted for this branch of work, and preparation for dinner, varies from an hour and a

half to near three hours. At one, or in some plantations, at two o'clock, the bell summons them to deliver in the tale (i.e. the total) of their grass, and assemble to their field work. If the overseer thinks their bundles too small, or if they come too late with them, they are punished with a number of stripes from four to ten. Some masters have gone as far as fifty stripes, which effectually disable the culprit for weeks … About half an hour before sun set, they may be found scattered again over the land … to cull, blade by blade, from among the weeds, their scanty parcel of grass. About seven o'clock in the evening, or later, according to the season of the year, when the overseer can find leisure, they are called over by list, to deliver in their second bundles of grass; and the same punishment, as at noon, is inflicted on the delinquents. Then they separate, in their way to their huts, a little brush-wood, or dry cow-dung, to prepare some simple mess for supper, and tomorrow's breakfast. This employs them till near midnight, and then they go to sleep till the bell calls them in the morning. (cited in Williams 1945)

The parallel with industry is striking. Popi and Lovely, who are fifteen and sixteen, work in a Dhaka garment factory. They live about three kilometres from their work, in a slum hutment beside the railway track. They are up each morning before five o'clock. They fetch water, wash and take breakfast, which is the remains of the previous night's rice to which a little water is added. They set out soon after seven to reach the factory by eight. They work as operators, machining shirts for export. Work starts at eight o'clock. There is a break at one o'clock when they eat chappati and vegetable which they carry in a three-layered metal tiffin can. Work starts again at two until six. Then a break of half an hour, and work until nine thirty or ten, according to the volume of work. Finally forty-five minutes' walk along the railway track, in the darkness, the white eye of a passing train occasionally cutting through the night with its long white cone, and a pause at the market to buy vegetable and rice for the evening meal. They go to Popi's mother, who has been working as a maidservant at several houses in the rich colony nearby.

Lovely's family remain in the small town in the south of Bangladesh, where Lovely's father is a cycle-rickshaw driver. On Fridays, work stops at two o'clock. The workers have a few days' holiday at Eid, but apart from that they work every day of the year. They are from Barisal, six or seven hours by bus from Dhaka. Their parents were labourers in paddy fields. Nothing prepared them for the relentlessness of life in the city; the accumulating deprivation of sleep, the tension and stress of fourteen hours a day in a factory.

Sometimes, they say, they think it will end. They will go back to Barisal, they will marry; only that would mean another captivity, the slavery of marriage, in a house from which their husband may never permit them to emerge. Popi and Lovely were not, of course, taken forcibly by slavers; but it is evidence only of the superior power of the impersonal forces which have removed them from family, home and village that the more corporeal ravishers and abductors of children can be dispensed with in the enlightened age of universal industrialisation.

Sometimes, the girls and women who work in the factories are beaten by foremen or overseers. If they make a mistake in their work, if their attention strays and the seams are crooked, if the machine becomes tangled with cotton, if their concentration slackens, they may be struck by an open hand, a stick or even the leg of a chair. After some years in the factory, they suffer from respiratory disorders, asthma, TB, loss of eyesight, and a constant cough. Few older women work in the factories. James Walvin (1993:137) writes:

> Even those slaves who survived tended to be dogged throughout their lives by recurring ailments and accidents. Register (later renamed Charles Grant), bought by Worthy Park at the age of eleven, began to work in the Third gang when he was twelve. After years in the fields, Register was shifted to the boiler-house – infamous for its intolerable heat as the cane juice was boiled and processed – possibly in 1816. Seven years later, he was described as 'diseased' at the age of forty-one.

In Bangladesh, much of the concern for working children comes primarily from NGOs (non-governmental organisations) and from international agencies. 'We are a poor country', say the rich of Bangladesh, gratefully finding in the poverty of the people a reason to resist redistributive justice. This view is echoed by the political parties – private wealth (which is considerable), far from being perceived as a scandal in the presence of such universal want, is considered to be insignificant in relation to total need and, therefore, scarcely worthy of being attacked, and certainly insignificant if it were a question of redistribution. The poor are brushed aside peremptorily, and the rich are inured to the most frightful spectacles of misfortune – the disfigured faces, the old women pushed by children in crude wooden carts, the truncated bodies propelling themselves along the road on a platform on wheels, the polio-withered limbs dangled entreatingly at car windows.

NGOs are often resented by government, for they are a constant reminder of its failings. NGOs provide schools, they offer training, they give health care. They are the contemporary equivalents of the reformers and social critics of Victorian England, the campaigners against the slave trade; with the difference that their non-governmental status distances them from government, whereas in Britain many of the reformers were also parliamentarians, and their work was often inspired by the findings of official Commissions of Enquiry, Royal Commissions and other formal investigations into the condition of the poor. While the government of Bangladesh will readily ratify the UN Convention on the Rights of the Child, they certainly do not have the will to implement any of its recommendations. This throws even greater responsibility upon the non-governmental organisations and, at the same time, often exposes them to official harassment and obstructionism.

I worked with two Bangladeshi NGOs involved in child labour which illuminated both the material similarities in the lives of working children through time and the cultural differences which reflect the influence of Islam and the traditional culture of Bengal at the end of the twentieth century.

Chapter Three

The Underprivileged Children's Education Programme (UCEP) was established in the early 1970s, soon after Bangladeshi independence. Although founded by a New Zealander, it has been thoroughly indigenised. (We should not be too concerned about the role of foreign involvement in NGO activity: for in this context foreigners are the equivalent of the upper-middle-class reformers in Victorian Britain who did so much to bring to the attention of the ruling classes the suffering in their midst. Doubtless leaders of the organised poor also emerged – much as has occurred in Bangladesh – but without an alliance with those who represented the conscience of the rich, they would almost certainly have been less effective, and possibly even more ruthlessly suppressed than they were.)

UCEP pioneered non-formal education, which is now generally accepted as the only way to reach children who work. Indeed, almost all NGOs working with children now offer informal education. The difference is that UCEP teaches children already working, and seeks to build up and enhance their skills for the future. Tanbir ul Islam Siddiqui says:

> Millions of dollars are spent on informal education. It is unproductive. They give three or four years general education, but have no idea of its future impact. They seek only to have large numbers of children pass through their system – the fate of the children in later life is not recorded by most of them.

UCEP combines general with technical education, articulated specifically to work. It also runs a job placement service, whereby young people can maximise the use of their skills and employers can find trained personnel. They are already skilled or semi-skilled when they come into the labour market. Placements are principally in the electrical trades, welding and fitting, refrigeration, automobile repairs, carpentry, textiles, knitting, spinning and weaving, garment finishing, tailoring, and factory work, producing TV sets and electronic goods.

No family is too poor to take part in the UCEP programme – flower-sellers, rag-pickers, brick-breakers, cart-pushers, vendors

– children involved in some of the most arduous labour take part
in it. There are three shifts daily in the school, so the facilities are
used intensively, and children can attend school without inter-
rupting their hours of work. UCEP achieves in four and a half
years what government schools take eight years to accomplish.

There is a process of selection involving the children, their
parents and employers. Boys are selected at eleven, girls at ten.
The families must be committed to their children's education and
promise that they will be allowed to complete the four and a half
years – that is up to grade eight. There are UCEP schools all over
Bangladesh, and the brightest children go on to technical school
in Dhaka and Chittagong.

Tanbir Siddiqui explained the conditions under which children
are registered.

> UCEP workers meet with the child and her or his parents or
> guardians, visit her or his place of work. We assess the child's
> contribution to family income, find out if they are part of a stable
> community, whether the parents will support them throughout
> their time in the school. The children are all working; although
> UCEP is now extending its scope, admitting those with disabil-
> ities and learning difficulties. There are in Bangladesh more than
> 10 million disabled people; the education of most of them has
> been neglected. In order to avoid the spectacle of the disabled
> standing in lines of wooden carts in front of the mosques on
> Friday, they must be provided with some skills that will enable
> them to participate in productive activity.
>
> The elimination of child labour remains our long-term
> objective. Our contribution to this is to prepare adults who will
> have the earning capacity to dispense with the labour of their
> own children. Even when the children have finished their
> schooling and have been placed with employers, we follow them
> up after six months and one year; if it doesn't work out, we find
> out why or we try to find them another job. UCEP is in contact
> with more than 1,000 employers – transnational companies,
> national firms, small industries and employers of a handful of
> workers. UCEP graduates are employed in all the leading
> companies in Bangladesh; they are preferred, because they are
> likely to be imbued with habits of industry and seriousness.

UCEP schools are advantaged by taking only children who are
likely to succeed. They do not take street children, those without
shelter, those without a settled way of life, since the dropout rate
among such children is high. The selective process suggests the
creation of new hierarchies within the poor themselves, dividing
families determined to give their children a better chance from

those who have not the will or the ability to support their children. The criticism remains – as with many other present-day remedies for poverty, including micro-credit – that their efforts simply redistribute between poor and very poor, thereby leaving grosser injustices intact.

The system operated by UCEP reminds me of the selective system, whereby able working-class children were privileged in Britain in our own time. Those in Britain who were given scholarships, and were subsidised to rise out of the working class through grammar school and university, recognise the limits of such systems: the divisiveness, the generation of snobbery and feelings of superiority in those who were chosen create different problems.

There are other imponderables in Bangladesh. Every country in the world is now talking of 'increasing the skills of its workforce' – from Britain to Thailand, from Brazil to India. What is the future of Bangladesh in an international division of labour which is intensely competitive, and in which Bangladesh starts with many disadvantages? If so many countries offer the labour of their people in a Dutch auction of lower and lower wages, of what use are the enhanced skills? Where national control is slipping away even from governments, how can the work of NGOs reverse or even halt processes which do not even originate in the country where they work?

I went to the Beribund school just behind Mohammadpur bus stand, along the embankment constructed to prevent flooding in the city, earth sticky from recent rain. Two or three hundred metres, and at the bottom of the slope, it is a long one-storied building, with a small yard in front. There are six classrooms, very crowded, and children are sitting in old-fashioned upright desks, extremely serious and attentive. Their faith in the instruction they receive is absolute. Again, not only the physical aspect of the place – chalkboard, desks which look as though they might have been shipped from my own school forty years ago – but the reverence of children for those who teach them are all poignant ghosts of a familiar past. I remember my mother saying with fervour how she loved her teacher: I used to consider this, and when I reviewed the weary and slothful men who taught me, I found the idea of loving them so implausible that I laughed dismissively. Here, in yet another personal illumination of our own past, I was able for the first time to understand what she meant.

In one class of fourteen- and fifteen-year-olds, there was a biscuit seller, a vegetable vendor, several brickworkers, a zari worker (embroidering sarees with gold and silver thread), some maidservants. Most of the fathers were rickshaw drivers, the mothers maidservants, although there were some garment workers and factory guards. One girl had left school to work in a garment factory, but her father had been persuaded to let her return. Half the children were born in the village and half in Dhaka. Here, the uprootings are still raw, city living seems improvised and provisional, even though few will ever return. This is how it must have been for the bewildered young factory workers in the Manchester of the 1820s. One girl is making handkerchiefs in a factory at 40 taka per shift (80 US cents). Most earn between 20 and 50 taka daily (40 cents and one dollar). One works with her mother as a maidservant. The majority work between six and eight hours daily. With their three hours' schooling, their duties at home (for education is often seen as a privilege, which they must pay for in other ways), it is a very long day. There is a serious absence of time to play. They need no instruction in social injustice, the gulf between rich and poor; they are also mindful of the difference between themselves and their parents, most of whom had little schooling. You can feel the disturbance this creates in the social order – the tilt in power from an older generation to a younger one: the children of migrants, even within their own country, are better equipped to deal with city life than their own parents and become interpreters, guides, as though industrial existence were also a foreign tongue.

The next time I went to Beribund, the rain started as I got down at the bus stand. Within a few seconds it had become a deluge. Looking for somewhere to shelter, I was approached by a man who offered me refuge in the office of his factory. The factory was clearly visible through gaps in the one-storey shops that shielded it from the road. There I caught sight of the most lurid image of the industrial inferno I have ever seen.

A glass factory: an uneven earth floor, a crude roof of corrugated metal supported by wooden and stone pillars. Rough furnaces with rounded tops, assembled in haste out of crumbling bricks, like two rows of outsize beehives, with other furnaces at random wherever there was space to build. These leave only narrow passageways for the men, women and child workers to move and work. The mouths of the furnaces gape with incandescent fire; the workers standing before them, flesh scarlet with reflected flame, stir the molten glass as though tormenting the food from the mouth of some mythical fire-breathing creature. Three men are working at one of these burning orifices, removing the molten glass on a long pole, dazzling gold orange spheres, which they plunge into a chemical liquid which will cool them slightly, so the globes change shape, dripping into an elongated oval. These are then poured into stone moulds, where they cool. They are then plucked out by young women and children with rough tongs, and placed in the pyramid of glass lamps in a corner of the factory, waiting to be packed and dispatched.

The people are close together, working in a rhythm designed not to bump into one another and so that the incandescent glass mass on the pole does not come into contact with the vulnerable flesh which itself seems to be illuminated from within by the fire reflected in it.

The whole building, which employs about 150 people, is alive with these spheres at various points of manufacture, so it looks like a planetarium – stars and constellations whirling around and creating islands of violent heat and light. The floor of the factory is covered with splinters and fragments of glass, a crystal floor covering that crunches beneath the workers' chappals. At the side of the furnace mouth, long fragments of glass that have fallen like saliva, as they drip from the pole, accumulate, delicate as slivers of spun sugar; a highly dangerous, beautiful formation of waste which will later be returned to the furnaces. At one fire, a man with a thinner pole is making small bottles of more delicate glass for medicine and perfume. A boy of about eleven is picking these up with flat metal pincers as they emerge from the mould; he places them in water so that the glass hardens in a mist of glazed smoke. He is earning 25 taka a day for this work. He throws misshapen bottles to one side so that they can be melted down again. A girl of about thirteen carries the finished bottles to the storage place.

There is no protection for the workers here: eyes, face, flesh – it is up to the individuals to make sure they do not injure themselves or each other. Their actions have a rhythmic precision, a rehearsed balletic movement; broken glass, molten glass, splintered glass, perfectly formed glass: only the people are more fragile, the bright dangerous orbs staining the naked skin gold with their heat and colour.

Many of the children at the UCEP school must pass this place each morning and evening. They do not have to look far to be made aware of their good fortune and to understand the inequalities which exist, even here, in this desolate place, which have granted them opportunities that the majority of children in Dhaka cannot hope for.

I went with Latif, one of the teachers, to visit the families of some of the working children. Shamina, who is twelve, is breaking bricks. She gets 3 or 4 taka per square foot of bricks (according to the size of the pieces). She will break about four square feet per day, which takes about four hours. In her family there are two girls and two boys. They had land in Faridpur, which was eroded by the river. Her father is a rickshaw driver and her brother a *ferrywala*. Her mother also works alongside Shamina.

Together, the family earn about 2,000 taka a month (US$ 40). Shamina's mother estimates that it costs 100 taka daily to feed a family of six adequately, so even if all income goes on food, there will be a deficit of nutrients. They never eat meat. Shamina's parents built the bamboo and *chetai* (panels of woven bamboo) house themselves. They are insecure here, the more so since a new road is being constructed, which will link Mohammadpur with the vast new monument being built to the martyrs of the Liberation War. This is a concrete semicircle on the edge of Beribund, where thousands of Bengali intellectuals, freedom fighters and activists were killed by the Pakistani army in 1971. This structure represents part of the effort by Sheikh Hasina's Awami League government to reclaim the moral credit for the freedom struggle from her rivals in the Bangladeshi National Party. That this monument will lead to loss of livelihood, displacement and insecurity of hundreds of families seems to count for nothing in those determined to celebrate dead martyrs, even if this means creating new ones out of the living.

Reba is twelve. She works four or five hours a day, breaking five or six square feet of bricks. She earns 18 to 20 taka a day. Her mother works slightly longer hours. Her father is a barber in a shop and earns 70 or 80 taka daily. Her sister, who is fourteen, works in a garment factory for 900 taka a month. Reba wants to go to technical school to learn zari work. The family came from Barisal. Landless, and in search of work, they settled here because people they knew had come before them. They pay 10 taka a month for water from the tubewell. Reba's mother says that the worst problem is the brick dust which covers everything. It gets into the food, it covers the bed, it is a constant irritant to the eyes and throat. They live in a bamboo and *chetai* hut, where a wooden bed occupies most of the interior. They pay only 200 taka monthly, but the family works on the land of the brick-owner in part payment for their accommodation.

The brickbreakers must supply their own tools, the hammers and stone anvils, and the protective rubber gloves that prevent hammer injuries. Reba gets up at six o'clock, washes herself and takes a breakfast of rice. She works until school-time, which runs from eleven forty-five to two twenty-five. After that, she takes lunch of fish or dal and vegetable, and then goes back to work. The family eat at eight o'clock in the evening and then sleep.

The school day has to be structured flexibly because, however committed a family may be to the child's education, the poor are often subject to emergencies which only their children can deal with – a young child is sick, there are floods, there is a problem with the family in the village. The children are part of the pooled resources of the family, and urgent business has to take precedence over education. In spite of this, attendance at the UCEP school is impressively high – close to or over 90 per cent.

We walked in the direction of the Martyrs' Monument which is about a kilometre from the beginning of the embankment. We came to rows of houses constructed on long bamboo stilts driven into the river bed. These are very crowded and reached by a rope and bamboo 'path' from the bank. There was a fire in these houses last year and seven people died. These structures are very dangerous – they may be blown away in storms, and a single spark from a cooking fire will destroy them in minutes.

Shahinur was just returning from work. A bright child of thirteen, she will almost certainly go on to technical school. She told us she had just finished her five square feet of bricks. Her earnings are 15 taka daily. There are three brothers and four sisters in the family. One brother sells fish and brings home about 50 taka a day. Her older sister teaches part-time in an NGO school and is paid 600 taka a month. Her father is too old to work.

Rabeya is Shahinur's mother. To reach their house, you have to go down the embankment, to where the earth has been flattened into a rough terrace. Even so, the house still tilts. Rabeya came from Barisal 'because of hardship' twenty-four years ago. They built their house themselves. The biggest problem is drinking water and dust. Children suffer from diarrhoea, fevers, chest infections and coughs. If they have money, people go to private doctors. In the government hospitals treatment is free, but your child may die while waiting for attention.

In Rabeya's family there are ten people. It costs between 125 and 150 taka daily to eat well. '*Mota-moti*', they eat so-so. They rarely have meat except at Eid; there may be mangoes or jackfruit at the height of the season. The house has been newly roofed. Rabeya took a loan from an NGO for 14,000 taka. She cooks outside when the weather is fine. The children pick up firewood where they can – to buy it would cost 15 or 20 taka a day. Rabeya says her family are 'moderately' poor. Five of her seven children are studying; she believes this will make their life easier as they find work.

The lives of poor families are a race against time: the children are there to help them with domestic work in the home, caring for younger children; by paid labour, and ultimately forming a shelter for the old when they are sick or too old for work. It is only natural that poor people should seek to shorten the period of time in which their children are dependent upon them. This accounts for the sad spectacle – often met with in Bangladesh – of whole families, including infants, sitting in their allotted spot on the embankment in front of a pile of bricks waiting to be broken, a little wall demarcating their space from that of their neighbour.

There is no doubt that the UCEP children are more easily employed than many others. More of them entertain the hope of going to technical schools than these can provide for. They make up a poor little 'aristocracy' of labour, since they have been disciplined and prepared for it from the age of ten or eleven. Naturally UCEP dwells upon its successes – whatever you do in Bangladesh, you cannot pause to consider what happens to all those who remain beyond the reach of your charitable endeavour.

Chapter Four

The NAYAN Foundation is an organisation named after the child of its founder, Abdur Rauf Bhuiyan, a boy who died of leukaemia at the age of three. Rather than simply grieving, his parents decided to celebrate his life by setting up a foundation which would help others. NAYAN works with children working in dangerous and hazardous occupations – the most difficult to help, and those who are presently the object of concern of the ILO and many charities in the West. Mr Bhuiyan says:

The acquisition of a skill in childhood, under whatever conditions, seems to the parents the most urgent priority, even if others derive even greater benefit from it. Employers make more profit if they employ children for very little pay. When I speak to the employers, they say, 'Oh, the parents accept it,' as though that absolved them from any responsibility.

The fundamental issue is, if you stop child labour, you stop the most important human right of all – the right to survive. If we provide decent pay for adults, they will not send their children to work. Employers say to me, 'If you take the children and educate them, do not bring them back to me afterwards, because they will know nothing of the work.' If you leave them where they are, within ten years, they will be managers of some small enterprise – automobiles, lathes, baby-taxi or rickshaw repair. The direct experience with the employer is as good as technical school.

They have a point. If the employer is good, the child is working no more than six hours a day, that is fine. The issue is whether the employers are exploitative, and whether the kind of skill they give is socially beneficial. Perception is everything.

You cannot understand Bangladesh without taking into account the influence of the global economy on our country. There are many good things in Bangladesh – in our social lives and family structures, our friendships and human commitment, we are stronger than the West; but you are trying to damage, to undermine, the good things we have here, without the benefit of your advantages. We are captive in a market mechanism we cannot control. People want to stop child labour here, but without any idea how it can be done, and certainly without paying attention to the inadequate rewards paid to adults. The issue is, when you are in an advantageous position, you can dictate to others. Those who come to invest, which is presented as though it were the salvation of the country, actually take much more than they give. There is a Bangla proverb, 'The wife of the poor man is the sister-in-law of everyone.'

The Harkin Bill was very damaging. What is a child? Western conceptions of childhood, which they defend so strictly, are themselves a fairly recent invention. Our borderline between childhood and adulthood is fluid; some children may be ready for work at eleven, others older, some maybe younger. Western categories are so rigid.

Observation suggests that there is a low dropout rate from small enterprises – if a child stays two years, he will usually continue. After two years they draw a salary. Education is an important intervention, but not the only one. Of course, education of the employers is also essential. Ask him why he is not putting his own son to work. When his humanity is appealed to, then he may look to the children he employs.

Child labour needs critical support from the employer, not from NGOs who have no responsibility for the children they 'liberate'. Employers are not nurses, but the human input into the work situation is the real education. Positive elements in child work have to be built on. Small entrepreneurs are as good as vocational training centres. At present some are exploitative, but they also serve to build careers for young people. You find no one with a technical skill sleeping in the street. If there are 6 million child workers, there are 600,000 employers investing capital that comes to 30 trillion taka (that is, 600 million US dollars) to train them. If we withdraw them, there are one hundred government orphanages run by the state, and a few hundred vocational schools. How will they accomplish a fraction of that work?

We have three targets, the children themselves, employers and the workers' families, to ensure that children get to participate in normal activities of childhood – play and education as well as labour.

We are talking of children who are not organised, who work on garbage heaps, as street vendors, hotel boys, cleaners and helpers, bringers of water, market porters, workers in garages, small manufacturing units. It has taken us three years to convince the employers to give the children two hours a day to attend school. We run it three days a week, because we do not wish to upset the employers too much. We can set these up anywhere –

at the roadside, under a tree, in a hut. School is not attendance-based, but learning-based, so it is flexible.

We have to look at the gradual elimination of child labour. If we destroy it through donors' conditions we shall do incalculable harm to those we are supposed to be 'rescuing'. Dhaka is full of reconditioned and repaired vehicles – repair shops come up like mushrooms. These repairs are our form of import substitution. No wonder the West wishes to destroy them. If we can make a new car, a new scooter out of one that is in ruins, it is magic for us, it shows the technical skill and ingenuity of our people. And this is learned by young children. It is a precious resource, and we have no wish to see it sacrificed for the sake of increasing imports of throwaway goods that cannot be repaired or recycled.

Employers have to be sensitised to the child's needs. Child labour is a growing force in our country. The employer makes a real contribution: the exploiter must become trainer, see himself as developer of the valuable human resources he has in his charge: the technical skill of employers allied to social education and time for play – development as human, not just material achievement. We want to produce in children the potential to be employed, not more and more of the educated or half-educated unemployed.

NAYAN has identified forty types of hazardous labour which create health hazards as well as physical danger – pillow-making and quilt-making which induce TB and lung cancer, working with chemicals, selling flowers among the traffic, operating lathes and welding. There are over 300 occupations in which children engage, of which forty are hazardous and twenty very hazardous.

I went with Anup, who works with NAYAN, to the industrial area of Doloikal in Old Dhaka. To get there we went by motorbike on an eventful drive through the city, knees scraping against truckloads of metal, being grazed by three-wheeler trucks in the gaseous streets of stationary traffic. At one point the road was blocked by an angry crowd. Thousands of demonstrators had besieged the Dhaka Electricity Supply Undertaking. For the past week, the power supply had been interrupted because of lightning damage. When the power supply fails, the water supply also breaks down, since this depends on electric power for pumps. Many poor areas have been without water for several days.

As we approach we can see the crowd ransacking the DESU building. They are stoning some trucks. A train is attacked. The offices are destroyed. We cannot pass; other vehicles are turning back for fear of being caught up in the violence. The past five days in Dhaka have been frightful. In the darkness, private generators throw their thin cones of light on to the stony potholed roads and their storms of dust and fumes, and the glistening bodies of sweating cycle-rickshaw drivers gleam in the headlights of passing cars. The workers on building sites illuminated by powerful arc-lights in a tableau of nocturnal labour, the stationary trucks farting their gases into the stalled traffic, the tangle of rickshaws, the nerves of overworked people jangling with heat, thirst and the perpetual haze of particles and fumes, have all contributed to the sense of urban confusion. The outbreak of violence is scarcely surprising; it is itself only a response to an immeasurably greater violence.

Instead of abandoning the trip to Doloikal, we go by a longer route. Doloikal is an area of small workshops and factories, and semi-open ground where all kinds of vehicles are repaired – cycle-rickshaws, baby-taxis, Tempos (these are small trucks used for passengers, smaller versions of Manila's jeepneys), motorcycles, trucks, cars. Partially dismantled frames stand haphazardly on the rough ground. Vehicles that would long ago have been abandoned in Europe are being reconditioned and made serviceable. There are sheds storing engine parts, spare wheels, mudguards, fenders, plugs, dashboards, clocks, panels and so on. Each small garage employs between about eight and fifteen people. Heads are bent over engines, four or five crowns of jet-black hair shining in the sunlight, the petals of a big dark flower. Pools of oil make rainbows on the uneven ground; a truck perilously balanced on its side while fragile bodies lie beneath it in defiance of gravity and their own safety. Small children are wrenching off wheels bigger than they are, mending tyres or bringing glasses of sweet tea in plastic crates. A scene of chaotic yet purposeful activity in which no hands are unoccupied.

On one side of this open space, a row of shops: suppliers of spare parts, as well as building materials and industrial tools. Crooked stone-built lanes lead off this road, tenements, dim alleys, narrow passages and low doors, a secret labyrinthine world that suggests violence, cruelty, crime. And indeed, all these things are present – economic violence, exploitative cruelty and crimes against the humanity of child workers. Some of the workshops are concrete chambers carved out of a portion of a building, only a few metres square, so full of machinery, there is scarcely room for the operatives to stand – polishing and grinding-wheels, metal-cutting machines, lathes and dies. Piles of cranks, cables, chassis, axles, wheels, some rusting, others polished and refurbished. Some children polishing metal are themselves covered with rust, so that the metal they handle is gleaming while they have the aspect of abandoned human sculptures.

In these units, perhaps one quarter of the workers are under fifteen. Most are between eleven and fourteen, but some are younger, many very small for their age; covered with dust, oil, grease, clothes stained beyond colour, faces grimy with the material of their labour. There are many unpaid apprentices. Some bring their own food – a couple of folded *rotis* and a spoonful of curried vegetable – in a tiffin can. Only the faith that they will learn a marketable skill binds them to the metallic clatter and chaos of the place.

This spare evocation gives no sense of the concentrated oppressive intensity. The humidity is itself saturated with violence; a violence that breaks out periodically when two cycle-rickshaws become entangled and the drivers come quickly to blows. The water-cart arriving after two days without water creates a tumult as children come running with plastic buckets, old paint cans, petrol containers. Two men astride the water cylinder pour it out at random and spill as much on to the dust as into the proffered vessels. Suddenly, after a twenty-four-hour break, the electricity is restored and the factories that depend upon it resume work frantically to make up for lost time; the workers, too, move as though they also were plugged into some invisible industrial power supply.

The energy and activity of the workers in the small units are overwhelming: cutting, drilling, hammering, beating, bending metal. A boy at a sort of mangle turns a handle that chops slices of rectangular metal for boxes and containers; another is drilling through slabs of metal for gearboxes, leaving corkscrew curls of silver. One is planing wood that comes away in pale brown flakes; another is pressing planks against a whirring wheel that produces a blizzard of sawdust which settles on his hair and eyebrows. Some young men are galvanising metal – razors, buckles, locks, bathroom fittings – holding them over a vat of chemicals through which a rod passes an electric current. With his bare hands a boy of fourteen steeps the component parts of scissors and knives, and blue sparks crackle and spit as he immerses each object for a moment. A child of about twelve plunges metal objects into acid, his only protection a pair of rubber gloves. Cramped, hectic places, some with brown water ankle-deep on the floor, with uncleared waste, dust covering everything, cobwebs hung with sawdust and grime, a thin light bulb on a bare flex casting the only light in windowless caves. Heat, semi-darkness, the smell of metal, dust dancing in an oblique wedge of orange sunlight entering an open door.

A shop where engines are reconditioned is where thirteen-year-old Nazim started work when he was ten. He receives no pay. He explains, 'I am still learning.' He works on truck engines. Nazim studied up to class three, until he was eight. He can write his name. The oldest of three brothers, he is confident he will start earning 'soon', although he is clearly already a valuable worker. He expects to get 200 taka a month (US\$ 4) as soon as the owner judges that he is competent. The employers are two brothers, one of whom explains that they set up the workshop ten years ago on what is government land. They started with only 3,000 taka and now they have capital of 300,000. Sixteen people work here, eight of them under fifteen. Those who are learning must expect to spend two years as apprentices, without pay. But it takes five years to become a skilled mechanic. Of the sixteen workers, six are unpaid. Nazim wants to become an employer. Since he lives nearby, he can go home for food in the middle of his twelve-hour day.

Roubel is in a vehicle repair shop. Also thirteen, he is one of three brothers and three sisters. His father is a cart-puller, his mother a maidservant. He earns 100 taka a week and was paid from the beginning, at the age of six, he thinks. One day, he also will become an employer. He works from eight o'clock in the morning until ten at night.

There is great variation in conditions and pay. All work long hours, but some are paid at least a small wage from the beginning. The boys do not complain, and there appears to be no jealousy, although they surely must wonder why some are so much better treated than others. Most are animated by a hope that they will one day become *malik* (owner) and will themselves be in a position to employ others.

Roton is twenty-two. He started in the truck repair unit when he was fourteen. He is earning 1,200 taka a month. One of four brothers, his father owns five baby-taxis, one of which he drives. Roton studied until class four, when he was nine, and he can write his own name. One of his brothers works in a clothing-machinery factory. Another of the anomalies in these units is the wide variety of economic circumstances of the families of the workers. If his father owns five baby-taxis, there was no economic need for Roton to leave school at nine. It seems that the *opportunity* for children to work is as powerful a factor as whether the money is needed or not. In the lives of many working children, there is a strong opportunistic element. The children of migrants sometimes work out of cultural habit – because nine-year-olds worked in the countryside, they should do likewise in the city. Perhaps money has its own seasons, like paddy which used to represent wealth – if it is not harvested when it is ready, it will not wait until tomorrow. Sometimes strange cultural continuities link the old life with the new.

Shopun is the owner of a small factory making concrete mixers for construction. The whole article is made in a very small space and spills over on to the road in front of the workshop. Shopun employs forty-two workers, ten of them under fifteen. All are paid, the youngest 600 a month. After four years they receive 3,000 to 4,000, and after ten years 10,000. The company was started by Shopun's father and now has a capital of 1 million taka. Employees work from nine to six in the evening, and after that they are paid overtime. Shopun is one of NAYAN's model employers. He is prepared to let the children receive education, although he is certain that literacy makes no difference to the quality of their work. This suggests that, from the point of view of the employers, time spent by the children in learning is not profitable. But Shopun can see that it is better for them and for society and that, in that sense, there may be some tangible benefit for him also.

Apu, fifteen, works in a baby-taxi repair shop. He has been here only two months. Before that, he worked in a factory making leather purses, but he was 'terminated' because he was late for work. He earns 500 taka a month, which he gives to his parents, apart from 30 taka he keeps for himself. There is one other brother and two sisters, and is father is a *mistri*, or mechanic, which Apu also wants to be.

Litton is in a small workshop which galvanises metal. It is very hot and there is one tiny window. The doorway is low, the floor uneven and wet. There are vats of chemicals, improvised loops of wire which transmit an electric current to a rod which Litton dips into a solution of copper, sodium and salt for the coating of buckles and metal accessories for furniture – mirror fittings and soap dishes. Litton is fourteen and earns 150 taka weekly. He is from Barisal, where his parents and family remain. In Dhaka, he stays with his aunt, his mother's sister. Family networks bring many children to the city – it is the channel for information, protection and finding work. Litton sends home 250 to 300 taka a month. He had no formal education. His clothes are corroded and stained by acid; he wears open plastic chappals. His working hours are from nine to seven daily.

In another concrete cell, two boys are washing and polishing galvanised metal soap dishes – shallow concave holders with a slit for the water to run off. They work at a kind of electrically operated grindstone. Metal dust is everywhere. Mazamul Haq is thirteen or fourteen – he is not quite sure. He wears a coloured handkerchief around his head to protect his hair from the dust, but nothing over his mouth or nose and no protection to stop the small pieces of metal from going into his eyes. He came from Mymensingh a year ago with the older boy, Faisal, who is eighteen and who has already worked in another place. They both work, sleep and eat here. Mazamul earns 800 taka a month, Faisal 2,000, of which he sends home 1,200 monthly. Mazamul has no parents. Both died when he was young, and he was living with an aunt, to whom he sends most of his earnings. The boys work from nine to six, and after that, overtime is paid at 10 taka an hour. Most evenings they work until nine or ten. Behind the polishing shop, there is a small space with two wooden bunks, a little stove on which a pot of rice is cooking, and a kettle. There is a change of clothing on a string and a plastic bucket for washing. Everything is covered with dust – even the visibility within the workplace is impaired by the tiny particles of metal. Mazamul sits at the grindstone, clearly unaware how damaging this dust must be to his lungs. It is not the employer who will pay the certain medical expenses which these young men's families will occur, expenses that will probably far outstrip anything they have earned through their punitive, captive labour.

In the centre of the rough ground, there is a shed of corrugated metal, which serves as mess and rest room for truck drivers whose vehicles are parked in the neighbourhood, waiting for repairs or for delivery of a load of material to be carried across Dhaka or to Chittagong or Khulna. Omar Faruk is a truck driver's helper. He is fourteen. Most drivers have a helper, but they are usually in their late teens. Omar starts work at five o'clock in the morning, and the journey often does not end until eleven or twelve at night. He earns 1,500 taka a month (US$ 30), two-thirds of which goes to his family: there are eight brothers and sisters. He lives in the truck, sleeping in the cabin. This, he says, pointing to the vehicle with its battered seats, is my home. His family owns thirty decimals of land, which they bought. They grow ten quintals of rice annually, not enough for self-reliance. The money he sends supplements the food they grow. His mother and father are in Rajshahi, the impoverished north. As well as looking after his own land, Omar's father is a share-cropper on the land of a neighbour. Omar has been in Dhaka for one year. Someone from his village recommended him to the truck driver who, he says, treats him well. As soon as he is old enough, he too wants to become a driver. His duties include loading and unloading, fetching food and water, guarding the vehicle when it is parked and doing elementary maintenance work.

Here are a few of the children whose employers NAYAN is trying to persuade of the wisdom of education. It is an uphill struggle. I talked one day with the staff, those whose job it is to convince the employers and to teach the children. The industrial areas of Dhaka have been divided up, and in each one there are an administrator and a teacher working.

Their principal problem illuminates the global problem at the local level. The children and their families join with the employers in a collusive resistance against what NAYAN is trying to do. Employers do not, in general, wish to lose labour time, and the children fear that if their work is interrupted, their job may be taken by others who will not bother with education. When the employers accuse NAYAN of interfering with their business, the parents tend to side with the workshop owners. These difficulties are reflected in the high staff turnover – few have been working for longer than one year. Only the most resolute will not be demoralised by indifferent employers. The children, for their part, want to learn, but earning is their priority. They have gained the recognition from some employers that childhood needs to be tempered by education and play; but some of the small owners say that they themselves started work as children, they had no education and they have succeeded. They are only helping a new generation to follow them.

Chapter Five

Many of the most conspicuous critics, of poverty in general and of child work in particular, are those working with foreign NGOs, people whose own countries have long ago pacified their poor by means of gradual economic improvement. Aware of the great material advances that have been achieved by the West, they come with the greatest goodwill to attain the same transformation in Bangladesh. It all appears straightforward: 'development' is a known path which all the countries of the South now wish to follow, indeed are compelled to follow in the absence of any alternative.

But those working for economic upliftment of the poor in Bangladesh, India and elsewhere are now pitted against vast transnational interests that are engaged upon deepening the extractive and impoverishing project which is globalisation. That this is now called 'wealth creation' and is supposed to unite rich and poor in joint dedication to its unquestionable virtues only obscures the unaskable question that hangs over all these poor countries – namely whether the means whereby the West got rich can be replicated here or not. And if not, what should they be expected to do?

Child labour – one of the most monstrous of many institutionalised injustices – is, perhaps, the greatest reproach to a model of growth and development that is now uncontested in the world. This is perhaps why it generates so much pietism and handwringing, so many words, so much jargon, so many convoluted defences, so many comforting evasions, so much talk now of gradualism, reform and improvement. For within a paradigm to which all alternatives have now been cancelled, it is only through the creation of much more wealth that the world can expect to see an end to work for the children of the poor. That this paradigm of perpetual increase is already striking against the limits of what the planet can bear is well known; yet it is still insisted that here lies the only hope and promise of deliverance. Child labour, perhaps more than anything else, is the most intractable challenge to the apologists for a future extrapolated from the present, for the destiny of every country on earth, inscribed in the progress and enlightenment of the history of the rich industrial nations.

In our time the argument between defenders of child labour and abolitionists is not presented as a conflict between the employers' right to use any labour at the cheapest rate they can get and the moral horror inspired by the factory system. The debate has been recast in ways that are more appropriate to the contemporary sensibility. So now at issue is the clash between those who see the Western model of a labour-free childhood as a necessary prerequisite for a civilised society and those who defend the right of children – including many children themselves – to work. The former, and this includes many trade unionists, say that the employment of children depresses adult wages, which makes people poorer and drives more children to work. They see legislation as the best means of combating it, as occurred in nineteenth-century Britain. Abolitionists tend to see work and education as incompatible. Defenders of child work say that, despite conditions that are sometimes dangerous and damaging, children want to work. It offers them a chance for self-determination and responsibility. It gives them a function. The problem is the absence of suitable work, not work itself. In any case, children learn more in a work environment that provides them with skills than in a school environment in which they are taught a reach-me-down syllabus from an archaic Western academic tradition.

Of course, these positions are often influenced by vested interests: the tenderness of Western abolitionists for Third World children may well conceal a protectionist intention – a fear that competition from goods made in poor countries by underpaid juveniles may throw their own people out of work, particularly since they are no longer at liberty to employ children. Similarly, all the forces of inertia and fatalism may well be marshalled behind a more permissive response to the labour of children – and those who espouse the right of children to choose may well also be unwittingly caught up in this version of conservatism: if children in Bangladesh and elsewhere are asked what they want to do, it is only natural that they will say they want to help their families, even though at the same time they will almost invariably say that they want to learn.

What is needed, argues Duncan Green in *Hidden Lives* (1998), is a closer definition of both 'child' and 'work'. Limited hours of labour, especially if a young person is acquiring a useful skill, is certainly not the worst thing that can happen to a child, although fourteen hours a day in a glass factory or working in a mine most certainly is. Some work is certainly better than not working, and the focus should be on eliminating the worst while improving the conditions of the rest. Green also advocates 'tackling the underlying reasons why children work, including poverty, and a poor education system'. This would mean attacking the whole process of globalisation and the universal market economy which underpins it. Not that there is necessarily anything wrong with that; but it should be clear what is actually meant by apparently innocent-sounding prescriptions for what 'we' must do: as soon as the first person plural is heard, it means that a discussion is coming up against the limits of the thinkable. A notional global inclusive 'we' suggests a common aim, in this case, to release children from the worst oppressions of labour. But this is reserved almost solely for pious adjurations, not for practical and determined efforts to abolish the causes of child labour – inequality, poverty, culturally determined attitudes towards children. The first person plural is never used when it is a question of how the riches of the world are to be distributed. Then it is a question of 'they', those who receive only a fraction of the value of their labour, or 'they', who live in unparalleled – and often fortified – opulence in defiance of the most frightful misery on their doorstep.

There are, apart from the 'rescue' of children from factories, at least two other major factors emanating from the West which have tended to draw significant numbers of children into employment during the past generation or so; one of them severely material, the other somewhat less tangible, but scarcely less influential.

First of all, at any given moment the number of poor countries which are undergoing 'structural adjustment programmes' under the superintendence of the international financial institutions contains a majority of the world's poor. Debt and structural adjustment imposed by the IMF as a condition for further lending have led to the elevation of the market and a reduction of state spending in every country where this has been applied. Those who have borne the brunt of cuts in welfare, education, health and nutrition – rarely is defence subject to the same stringency – have been, of course, the most vulnerable. One consequence of this has been even greater social dislocation – more broken families, more working women, more children forced into the labour market.

The social costs of 'economic adjustment', as countries integrate themselves into the global economy, have been enormous. As though the 'normal' process of growth did not already create further inequality, the draconian necessities of structural adjustment have further exacerbated the position of the poor. They have suffered disproportionately as countries have been compelled to dismantle some of the protection they had sought to provide against the ravages of the free market. In Nicaragua, for instance, three-quarters of the people were below the poverty line in the wake of the free market policies conducted by Violet Chamorro after the overthrow of the Sandinistas. More children have been pushed to the margins by these developments. In any major city, from Manila to Mumbai, from Nova Iguacu to Lima, more children must scavenge on the rubbish tips, more children enter the sex trade, more go into the carceral factories of Jakarta or Port-au-Prince to make goods for export to service a bottomless debt.

The signatories to the United Nations Convention on the Rights of the Child in 1989 (which now includes almost every country in the world, notable exceptions being Somalia and the United States) readily abrogate their commitment in the name of economic necessity, a necessity imposed by institutions run, funded and dominated by the rich countries. This makes much of the pious devotion to the rights of children enunciated by those same countries mere theory. What they were actually doing throughout the 1990s was worsening the position of the poorest, thereby undermining children's chances of survival and forcing them into labour at an earlier age in order to do so.

The second pressure tending towards children entering the labour market is not so evident, but possibly even more compelling. It has been less commented upon than the structural adjustment programmes, but it is as pervasive as it is obvious. This is the global reach of consumerism, the spread of an iconography of luxury and affluence, and all the objects associated with a way of life which is promoted by the West as indispensable to a fulfilling human existence. This was initiated not only by a desire to create new markets but also as part of a competitive struggle with socialism, an endeavour to demonstrate to the poor of the earth that capitalism and not socialism held the key to all that was most desirable, that the good things of life were to be had as objects of private consumption. With the death of the Soviet Union, there has been no good reason to curb this message to the peoples of the world. Quite the contrary. It has all been stepped up, so that now the logos of transnationals are more recognisable than alphabets as they burn the night sky above the world's cities.

One consequence of this sanctification of the market has been, unsurprisingly, that human beings come to emulate the commodities which have been so intensively promoted in the world. The trade in children and women for the sex industry is only one aspect of this, as illustrated in the bleak transactions recorded by Siriporn Skrobanek in *The Traffic in Women* (1997): 'Saeng, a fourteen-year-old girl from Burma, was sold by her elder sister for 2,000 baht (US$ 80) ... A girl from the Akha hill tribe was sold for 2,000 baht when she was ten. She was later sold to a brothel in the south for 40,000 baht (about US$ 1,600).' Although the examples are rare, there have been cases of parents in impover-ished parts of the north-east of Thailand who have sold their daughters for the price of some coveted consumer object – a TV set or a refrigerator. It is not that the daughter is seen to be of equal worth to the value of such things, but the *symbolic* value of an item that appears to provide access to a new and more hopeful world than that bounded by rice fields, sky and poverty.

There are less spectacular instances where children work, not because the family is poor but because their labour will make the parents' lives easier. If a father is addicted to alcohol or drugs, children may sometimes be compelled to labour to provide for his needs. Sometimes, the cultural tradition which sent children to work in the fields as soon as they could be trusted with minding buffaloes or collecting fodder continues in the urban environment, even though there it may well no longer be an absolute economic necessity.

But the great majority of children work because their families are poor. The truth is that the parents of most working children depend upon their earnings to buy the food that will restore the energy that will permit them to work another day. They are not transfixed by materialism, they are not greedy or selfish – they simply see no other pathway to survival; and the children absorb this lesson in the daily experience of empty bellies and frequent sickness.

Doubtless, the ratification by so many governments of the 1989 United Nations Convention on the Rights of the Child marked a new departure for child protection. This engendered a certain euphoria about 'universal' commitment to the protection of the world's children, despite the fact that many countries of the South do not have the infrastructure to supervise the aspirations enshrined in the convention. If we add to this traditional cultural hierarchies characterised by indifference of the rich towards the poor, we can see more easily how the will of the 'global community' to protect children is weakened, if not cancelled, in its application.

The convention contains articles that have a direct bearing on child labour: Article 32 states the right to protection against economic exploitation and Article 28 covers the right to education. One of the problems with the convention is that it takes an idealised Western norm of family as the basis for the protection of children. Not only does this ignore extended and joint families of the sub-continent and elsewhere, but it promotes a version of the Western family which is already at an advanced state of dissolution. One of the issues that will emerge from this book is that there are other agents than poverty which subvert the autonomy, the peaceable growth and development of children. One of these is precisely what is regarded as the 'normal' environment for children in the West, namely that they should grow and develop, come to maturity under the ubiquitous primary determinant of the market, as though this were a force of nature and not a human-made structure which has its own necessities, its own internal dynamic, its own capacity for deforming human purposes.

In any case, there are other questions raised by the universalising thrust of the convention. It is now taken for granted that

'education' means state provision. The overwhelming image of
education in Bangladesh and India is a sterile classroom of battered
desks, few and archaic textbooks, a chalkboard, and a droning
teacher who lacks energy or commitment and, indeed, who in poor
areas frequently does not turn up to teach at all. Who is to say that
the children who emerge from five, seven or even ten years of such
a process are more educated than the illiterate Adivasi children of
the forests? Too many of the assumptions built into the statistics
and the measures of development or progress predetermine the
future of everyone on the planet by compelling them into a narrow
and culture-specific definition of what childhood, and education,
should be.

When I was working with Winin Pereira in Mahrashtra, in India,
we met a twelve-year-old lively and alert child who knew the uses
of hundreds of shrubs, trees, flowers and roots in her environment.
She knew where to find medicines against diarrhoea, leaves that
would stop bleeding, how to locate cures for snake bites, fevers,
swellings and sickness. She could identify fruits, roots and leaves
for food. She knew where to find fodder, which wood had a high
calorific value for cooking, how to make ropes from creepers. She
knew how to build a cool weatherproof house. She knew where to
find oil seeds that could be used to make a slow-burning lamp.
She could prepare the rice fields before the monsoon and catch
fish by using powdered bark that stunned them as they swam into
her hands. She could trap rabbits, quail and other birds to
supplement the family diet. She wore an old cotton dress discoloured by age; she went without shoes.

She had not been to school, but she was the very embodiment
of a considerable intelligence which had been cultivated by the
environment in which she lived. Yet in the tables of the United
Nations Development Index, her illiteracy would register as
'backward' (see Pereira 1992). The same was, until recently, true
of the Chakmas of the Chittagong Hill Tracts in Bangladesh. The
aetiological myths, the stories and songs of the traditional ballad
singers filled the minds of the young, who had a phenomenal
memory span and who had complete knowledge systems that
permitted them to coexist easily and symbiotically with the deep
jungle environment. The destruction of the forest, the re-planting
with exotic commercial species, the loss of biodiversity, the ruin of
the ecology also demolished the knowledge system; in the words
of one Chakma elder, this was to the hill people a violence as great
as it would be 'if someone had destroyed all the books produced
by Western culture'. As it is, there is nothing left for their children
to do but to become pupils in the bleak confinement of concrete
classrooms, hoping, at the end of it, to find some lowly job in
government service.

Chapter Six

Bangladesh is not the only country to have come under the abolitionist threat from the West. Not only governments, but the media have also been busy 'exposing' abuses of child workers, particularly in factories which make goods for export to the West. One notable case of this was the revelation by a British television programme in 1995 that young girls between the ages of twelve and fifteen in Meknes, Morocco, had been making pyjamas for a supplier to Marks and Spencer in the UK. Many of the girls were dismissed, even though their families had thought this a perfectly acceptable and indeed, essential, occupation, which augmented their otherwise inadequate income.

Examples from all over the world suggest that the banning of child labour often plunges families into deeper impoverishment. Duncan Green, in *Hidden Lives* (1998), tells how near Recife in Brazil sugar cane cutters under the age of fourteen were banned after a trade union campaign against dangerous and exploitative conditions. Their families fell into worse impoverishment and malnutrition as a result. Moral crusades are often decontextualised, the consequences for their 'beneficiaries' are frequently not taken into consideration.

It is widely believed in the South that such pressures have more to do with the protection, not only of Western industries, but also of the conscience of Western consumers who increasingly do not want goods thought to be tainted with the sweat and exploitation of children. That the fate of the children disemployed by Western conscience was not considered demonstrated – as did the Harkin Bill dismissals – that the agenda of the West was not necessarily the well-being of the children themselves.

Nor are these the only examples. When extra-territorial legislation which would allow them to pursue their own nationals who had abused children in the Third World was enacted in many Western countries, the concern was essentially to show the world that the West did not tolerate this kind of behaviour. Of the cases which have so far come to court, what happened to the children who had been abused was, in almost every case, unknown: they disappeared into the streets and slums from where they had been plucked by their abusers. Street children, prostituted children and homeless orphans became of interest simply because they had been the prey of sex offenders. They could be taken to Australia or Sweden, lodged in five-star hotels, produced before the court to give evidence against their abusers, only to be swallowed up in the night of unknowing from which they had briefly emerged for the chastisement of the wrongdoers. Once again, it was the preservation of the Western image, Western interests, Western world view that was at issue. Indifference to the destiny of Third World children – ostensibly the object of the exercise – occurs too frequently to make this simply a coincidence.

Although people are always poor in the same way, and want and hunger torment the body of Hindu and Muslim, animist and atheist, Nigerian and Burmese without discrimination, there is no doubt that different cultures and traditions inflect poverty differently. It is quite clear that the poor of Bangladesh are different in significant ways from their counterparts in Victorian England; and they are not the same as the poor of, say, Latin America or the Caribbean. Indeed, given the depth and intensity of Western influence on South America, the urban poor of Brazil or Guatemala resemble more closely their predecessors in the industrial Britain of the nineteenth century.

For one thing, levels of social breakdown in Brazil and parts of the Caribbean resemble the process of urbanisation as it affected the north of England in the early nineteenth century. The transformation of Brazil from a predominantly agricultural to an overwhelmingly urban society was a breakneck process – most of the change took place in the twenty years from the 1960s to the 1980s. It is only to be expected that social disintegration will affect some of the least defended. People simply cannot keep pace with the speed of the unchosen changes thrust upon them. The mental dislocation, the wounded psyche, the damaged humanity are all very conspicuous on the streets of Sao Paulo and Rio. The loss of an extended family, the consolations of alcohol and drugs, the destruction of rootedness and belonging, the absence of social constraints on new communities of migrants – all this was characteristic of the 'new' manufacturing towns of the north of England, the raw red brick of back-to-back housing which was the equivalent of the shanties of tin, wood and polythene that have grown haphazardly on the periphery of the South American cities.

The moral panics which have swept through Brazil and other parts of the continent over criminal street children are a more vicious and destructive version of the more restrained agonising over the disreputable poor in the cities of Victorian Britain. I remember in the early 1990s visiting the spot where a number of street children had been murdered, in a place called Nova Gerusalemme outside Sao Paulo: the banality of the setting – a dusty concrete bridge over a railway track on which people were desperately selling a few nuts and bolts, combs, toothbrushes, sweets, a handful of bananas, some oranges, second-hand shoes – only underlined the senseless cruelty of the cultural cleansing that saw 5,644 children between the ages of five and seventeen meet a violent death in Brazil between 1988 and 1991. The unwanted children of early industrialism in Britain found their way into the parish orphanages, and these were then transferred to the new factories of the north, where they were often beaten or worked, not infrequently, to death. The only difference is that Britain exhibited a more systematic and utilitarian method of dealing with the unwanted orphans of industrialism, the starvelings of violent social upheaval. With characteristic puritan rigour, we put them to work.

Bangladesh has not been subjected to the same dramatic transformation seen in Brazil. It remains substantially agricultural, with almost two-thirds of its people still employed in the agricultural sector. In Bangladesh the children are not 'cleansed' from the streets, except occasionally by death from 'natural' causes (that is, neglect rather than extermination). They beg, they wheedle, they cajole, they touch the heart. They do not form a violent or criminal subculture. They may be a reproach to the rich, but the rich of Bangladesh are used to such spectacles and readily shake off the importunate entreaties of street children. They have accommodated themselves to the poor wisps of humanity who live, and sometimes die, on the sidewalks, over whose huddled bodies they step as they go about their untroubled errands in the city. No one seems to think they show up Bangladeshi society in a poor light, and therefore suggests they should be eliminated.

The Indian subcontinent has its own cultural reasons, its own long history of acceptable social suffering, which urbanisation has only reshaped and 'modernised'. Such reassuring continuities are absent in Brazil, that country where so many freed slaves made their home and where, today, we may see a telescoped version of the worst of early urbanisation together with the renewed savageries of Western society in its post-industrial incarnation, with its drug-dominated slums, its gun culture, its gangs, its addictions and turf wars over some of the most derelict landscapes ever laid waste by human societies.

Chapter Seven

The images and descriptions of an earlier Western experience are sufficiently powerful to create the impression that 'development' is, after all, a predictable, almost predetermined pathway. As part and parcel of this, it is taken for granted that, if the creation of wealth is allowed to take its 'natural' course, all known evils will be swept away. And this is the unexamined core belief of the prescriptions, remedies and policies foisted by the powerful upon the poor of Bangladesh.

And indeed, so great are the similarities that it is sometimes difficult to distinguish whether reports on the work of children are referring to London in the 1850s or Dhaka in the 1990s. E.P. Thompson, in his *Making of the English Working Class*, writes of the break-up of the family economy, which led to the intensification of child labour between 1780 and 1830.

> In all homes girls were occupied about the baking, brewing, cleaning and chores. In agriculture, children – often ill-clothed – would work in all weathers in the fields or about the farm. But, when compared with the factory system, there are some important qualifications. There was some variety of employment (and monotony is particularly cruel to the child) ... We may suppose a graduated introduction to work, with some relation to the child's capacities and age, interspersed with running messages, blackberrying, fuel-gathering and play. Above all, the work was within the family economy and under parental care.

When the Hammonds wrote about the cumulative impact of enclosures, the agricultural revolution and the destruction of the commons, their descriptions anticipate the fate of the rural population of the contemporary world.

> The transformation of the village brought new problems. The peasant had partly fed and partly clothed himself (sic). His place had now been taken by the labourer who depended on the farmer for wages and the shop for food. By a paradox that roused Cobbett to fury, the village itself went short of food under the system that was maintaining the rapidly growing population of

Manchester and Leeds. The farmers, producing for a larger market, would not take the trouble to supply their labourers with milk and the milk went out of the village, so that the labourer who no longer had a cow could not get milk for love or money. The price of his flour went up because the farmer now sold to the miller, the miller to the mealman and the mealman to the shop ... It was a paradox of the new system that agriculture became very profitable, and yet the mass of the workers in this thriving industry sank into greater poverty.

Their distress – which erupted into riots in the South of England in the winter of 1830 – drove more people towards the industrial centres of Birmingham, Bristol and London. The parallels are striking in present-day Bangladesh; the only thing missing is the militancy and the uprisings. In fact, the Leftist militants of the 1950s and 1960s were ruthlessly suppressed after Bangladesh became independent in 1971, with the result that many activists fled to the forests, where they degenerated into criminal gangs, dacoits and terrorists, their original purpose of social justice degraded into armed robbery and plunder.

The emergence of children from rural or semi-rural family life into the factories was always a terrifying experience. P. Gaskell in *The Manufacturing Population of England* (1833) saw that the workers' discontent arose less from simple wage issues than from 'the separation of families, breaking up of households, the disruption of all those ties which link man's heart to the better portion of his nature – viz. his instincts and social affections'. That people 'chose' to move to the towns, or that men 'elected' to migrate for work or children were indentured to employers in distant places, makes it no less violent a transformation than those seized by foreigners and taken by force through the rigours of the 'middle passage' to the islands of the Caribbean.

The shock of change is not quite the same in the South today, even though the country sensibility is under a similar assault, the destruction of the peasant psyche is no less violent than it was then. But the bonds of family and kinship remain: the young workers, almost without exception, say they are working for the sake of their parents, their brothers and sisters. Furthermore, they regard work in the factories as a symbol of entering the modern world. Their counterparts in Britain in the 1820s did not know at that time on what kind of a journey they were embarking. They did not have before their eyes the universal iconography of wealth associated with the global advertising and publicity industries, the imagery of worldwide entertainment conglomerates, a pervasive picture of affluence and ease.

Today the transition is likely to be made with less reluctance, even though the impact of the reality may be just as traumatic. Indeed, migration to the city is often a profoundly disturbing experience, whether for factory work or domestic service. Therese Blanchet records the story of a mother who lost track of her daughter when the family with whom she was serving moved without informing the mother. In another instance, a boy domestic locked in the house where he worked was permitted to speak to his mother only through the bolted door. The sundering of families was also one of the cruellest aspects of the slave trade, made worse by the conviction of those whose business it was that the slaves 'were strangers to all normal human affections', and that they would therefore be exempt from the expected emotions at being wrenched from the home place and those they loved.

The literature of the nineteenth century in Britain was full of disappearances, random separations and coincidental reunions which often gave a satisfactory close to the novel, but were often criticised as sentimental or implausible. In view of the driven, turbulent forces which disrupted the lives of the people scattered and drove them apart, the Gothic plots of Victorian novels may be less fanciful than they appear to posterity.

Of all the employments which give work to children in Dhaka, few are more. desolating than the making and breaking of bricks: it sometimes seems that the whole city is a place of Sisyphean labour, where, for some cosmic existential wrong they have done, people are condemned to an eternal labour of making and breaking bricks; an endless punishment with neither meaning nor purpose. It is not so much the physical aspect of nineteenth-century mills and factories that is recreated in Dhaka as the sense of the repetitiveness of the labour performed in them. Just as a family likeness is perceived, not in the precise features of a relative, but in an expression that passes over the face from time to time, so it is not mimicry of identical labour that characterises much of the labour in the cities of South Asia today so much as the unvarying sameness of the actions required to perform it.

Beribund is a causeway on top of a dyke constructed to the north-west of Dhaka to protect the city from floods. On one side is a polluted pond choked with water-hyacinths and surrounded by concrete buildings and middle-class flats; on the other, a tributary of the river that brings country boats from the brickfields. There are paddy fields on either side with a growth of emerald green, and some streams meandering from the river, which will themselves become rolling waterways when the heavy rains mix with the meltwater from the distant Himalayan glaciers and finally reach the Bay of Bengal.

The embankment is about ten metres above the waterline in February, the dry season. In the sweep of the dyke is a semicircle of slum huts: the bamboo walls are grey and wasting, bleached by the sun; while the privies – little platforms on bamboo stilts and surrounded by a length of jute sacking – trickle their dark waste into the water below. Between the houses are narrow walkways, less than half a metre wide. The houses are made of bamboo staves with panels of *chetai*; roofs of palm-leaves and polythene, occasionally corrugated metal.

Between the houses, on the ledges formed by the gentle slope of the embankment, people are breaking bricks. Whole families sit, sometimes beneath a collapsed umbrella, at stone anvils, tapping, a variable music of metal on stone, since the bricks are broken into pieces of different size – some about fifteen centimetres square, others of about ten centimetres, while the dust, which gathers in drifts around the workers, is collected separately. This will be mixed with cement and used for terraces and flooring in the new apartments that are going up all over Dhaka.

Over everything there is a film of dust: the women and children crouched over the piles of brick; a reddish smoke of dust rises from each anvil, settling into the folds of the sarees of the women, changing the skin colour of the children.

The causeway, about five metres wide, goes on for several kilometres. There is room for trucks bringing new bricks and taking broken ones to pass. The trucks themselves set up choking dust storms as they go by. Brick dust is as ubiquitous as the fog in the beginning of Dickens's evocation of London in *Bleak House*. The dust forms in the folds of polythene on the leaves. It discolours the palms. It turns the leaves of the occasional trees into rust-coloured metal sculptures.

Many children sit with their mothers and older siblings as they work, playing with pieces of brick instead of toys. The object of their future labour casts its shadow over their infancy. A child of about four sits working a small hammer on the friable brick, rhythmic, automatic, as though breaking brick was a human reflex, like eating, breathing, shitting; a premonitory disciplining by these relentless fields of brick. Is he playing or working? He is quite naked, his body covered with dust, his tiny cock almost lost in the mounting drift of dust that piles up around him.

More bricks arrive, more fragments depart. The people in this circle of hell seem to have been invaded by an alien element. A strange human-made destiny the colour of dried blood, of a setting sun, a landscape of rust of which a child, pausing for a moment in her work, seems to be for a moment a surreal emanation. Another child is crying: damp patches on her cheeks restore the skin to its natural amber. Back-breaking, brick-breaking, heartbreaking necessity. A labour against which no campaigns, no humanitarian movements have been started. Let them eat brick. They do anyway.

In the evening, the sun is less fierce, and some of the people are stopping work. Baby Begum sits with her daughters in her hut. A bamboo pole in the middle bears up a flimsy roof of broken grey bamboo. A wooden bed fills three-quarters of the room. There is a bedroll, a quilt, some children's clothing on the wooden bamboo frame, and a bamboo cupboard with cooking vessels and plates. Some clothes hang from a piece of string tied to the wall. A concrete floor slopes towards the door, which is in fact only a piece of coarse cloth. A bare electric bulb on a flex, a rusty ceiling fan. And dust over everything.

Baby Begum is in her late twenties. She was sent here with her uncle from their home in Barisal, where her parents had no land. Her father was too sick to work. Baby Begum had no schooling, but she can write her name. She began work at the age of five or six, breaking bricks. She now works with her daughter, breaking bricks for 50 or 60 taka a day (just over one dollar).

Baby Begum has three daughters and one son. Her husband is a rickshaw-van driver, that is, he drives goods on a flat cart behind a cycle-rickshaw. He takes vegetables to market and earns about 700 taka a month (US$ 14). He assembled the vehicle himself from the spare parts of a discarded vehicle. Baby Begum's daughter, Rubia, is eight. She works every day. In the morning she goes to a school run by UCEP in this area (see above). Every day after school she helps her mother.

Baby Begum and her husband bought their house for 4,000 taka. There are rumours of evictions. The Awami League government is building a monument to the martyrs of the independence struggle. The site of the monument is Raybazar, where the Pakistanis killed many Bengali intellectuals. The building which is under construction looms menacingly over Beribund. It is a huge structure, and it is said that a new road to reach it will wipe out much of the Beribund slum.

Shortage of drinking water, lack of sanitation, the implacable dust, breathing disorders and diarrhoea are constant problems. At one time, there were *mastaans*, thugs who used to terrorise the people and extort money, but that has now ceased. Baby Begum hopes her daughters will have an education. She will sell her blood to get money for her children to be educated. She will not let her oldest daughter, who is thirteen, break bricks, because she does not wish to damage her marriage prospects. If it is known that she is a brick-breaker, she will only be able to marry a rickshaw driver. If she is not working but has some education, she will marry someone 'with a job'; that means one who is not self-employed but works in a company or a factory. She says, 'We are uneducated, that is why we break bricks.'

The electricity is taken from an illegal connection which an enterprising man in the slum sells for 100 taka a month. The house itself is in need of repair. The bamboo is thin, there is a hole in the roof.

Baby Begum belongs to a micro-credit scheme, Safesave, which is distinguished from many other slum credit schemes in that people are not bound by a timetable determined by the convenience of the organisation. They can pay in or take out money on any day they choose. Baby Begum has saved more money in her daughters' names than in her own. When we ask why, she says, 'Doesn't my daughter have a future? Will she not be married?'

The bricks come from the kilns just beyond the city boundary. The highway out of Dhaka is a narrow road, choked with buses, trucks and Tempos. The traffic sets up permanent clouds of dust which blanch the landscape, discolour the vehicles and alter the colour of the crops.

The sulphurous landscapes evoked by Engels in 1844 are recreated more than a century and a half later here in Dhaka. A new road is under construction beside the old country road. The marshy land has been excavated for clay for the brickworks, the countryside is ravaged. The metal cylinders of the brickfield chimneys gleam in a sunlight made hazy by the smoke which travels in long vaporous ribbons of grey and black, staining the sky with their parallel furrows.

The brickworks are dense, with hardly any space between the individual fields. Rough causeways enable the trucks to turn round as they come off the highway. At the entrance to the brickyard a booth is guarded by a supervisor. Then mounds of coal, charcoal and wood. After that, a slope leading up to the plateau of rubble, stone and dust which seals in the furnaces while the bricks are baked and where perpetual fire burns at a very high temperature. About eighteen men are working here. They wear asbestos-soled chappals, so if they place a foot wrong on the hot surfaces their feet will not blister; but they are supposed to keep to the narrow pathways between the furnaces.

Just beyond the edge of the furnace, there is a roughly-built brick shelter, where about ten men live and sleep: a bare rectangle, with bedrolls, a loop of string from which a change of clothing hangs, a small bundle of possessions in a metal locker – the most austere sketch of a living place for men, seasonal migrants who arrive in October from the countryside and remain until the rains dampen the brickfields in April or May. During the working season, the present construction boom in Dhaka has created such a demand for bricks that they work every day, seven days a week during daylight hours; a labour of five or six months without remission, without holiday, without rest.

The subterranean fires are fed through a series of small apertures on the surface, covered by a round metal lid through which the smoke nevertheless escapes all the time in shimmering coils. As soon as the firemaster raises the lid with a metal pole, a gust of immense heat and the scarlet-white glow can be seen and felt. To become a proficient firemaster requires five or six seasons since the firing of bricks requires a precise temperature so that they are neither too friable nor semi-melted by overheating. The firemaster is a dark-skinned man of about thirty; he wears a fuel-stained shirt and *dhoti*. On one of the metal lids the evening meal is being prepared; a black pot of rice is bubbling vigorously. The men prepare their own food, which is high in carbohydrate, since the work requires much energy. They use canal water for washing, and drinking water comes from a tubewell.

The unbaked bricks of grey clay are piled in long rows behind the furnace. The firing is a continuous process: as one area of the kiln cools, the finished bricks are taken out and dispatched. Then another load of unbaked bricks is assembled in a new architecture before being sealed in the heat from the furnace. Each truckload of bricks is worth 5,000 taka (US$ 100). At each firing, between 300,000 and 500,000 bricks are made.

Most of the men are young. They cannot continue into middle age, although there is one veteran of perhaps forty, with a greying beard. Abdul Aziz is from Faridpur. He is just fifteen and has a darkening down on his upper lip, but the face of an adolescent and the feelings of a child. For him six months is a long time, and when he talks of home his eyes become moist. He wears a blue-check *lunghi* tied at the waist in a big knot and a faded burgundy shirt, unbuttoned to show a herringbone of hair on his smooth chest. He came here for the first time last season for one month and then returned in November for the season. His parents, brothers and sisters are at home, but they have no land and only occasional work. When he looks at the utter desolation of the landscape, he bitterly misses his family and village. There, everything is soft, the earth and growing things; here everything is hard, unyielding brick. That the natural elements of earth and fire should have been combined to create such an infernal place testifies to the ingenuity of human beings in our capacity to degrade not only the earth but other human creatures at the same time. Abdul Aziz earns 1,800 taka a month (US$ 36). Food alone costs 1,000 taka. He sends 200 taka to the impoverished homestead, a miserable sum, but which goes some way to sustaining the place from which poverty has wrenched him.

Talking with Aziz, I was reminded of Walvin's account of the induction of slaves into work on the plantations of the sugar islands.

On the plantation the newly acquired slave was put through a process known as 'seasoning' to adjust him psychologically to work as a slave and to teach him the tasks and routine of the plantation. During this period of about a year, which might however last as long as three years, a surprisingly large number, unable to make the adjustment, died ... The practice was to assign an already seasoned slave, preferably one who spoke the newcomer's native language, to assist in the task of breaking him in. This would include *endeavouring to overcome the state of melancholy and despondency to which he might have been reduced by his traumatic experience,* demonstrating the use of agricultural implements with which he might not be familiar, and advising him on the cultivation of the plot of ground which would have to provide the greater part of his food.

Chapter Eight

Official indifference to the fate of working children in Bangladesh has been noted. Indeed, corrupt and rapacious administrations in the South are now routinely deplored by Western governments and institutions. This appears to offer a sharp contrast to nineteenth-century Britain, although we should not necessarily believe that the busy-ness of the state was always effective in carrying out the investigations into the plight of the poor which it so ostentatiously set up. E.P. Thompson, writing of the members of the Factory Commission of 1833, says that they were criticised for wining and dining with the mill owners, and for spending only a derisory portion of their time on inspection. It was noted that mills were specially whitewashed and cleansed, and underage children removed from sight before their visits. This has its direct parallel in contemporary Dhaka. When those monitoring the effectiveness of the Memorandum of Understanding (under the provisions of the Harkin agreement, which was to have emptied the factories of child workers) arrived in the factories, the child workers were often whisked out of sight. In one factory I visited in 1998 – to which we were admitted with great reluctance, and only after an assurance that we were potential buyers from UK – we were just in time to see the door closing on the last of the underage workers being shut in the toilets.

The flagrant errors and misunderstandings over the Harkin Bill gave rise in the 1990s to more considered versions of what should be done about child work. Declarations, resolutions and conventions were in themselves no more effective than the passing of legislation in Western countries outlawing goods made with child labour. The most recent ILO Convention on the abolition of the worst aspects of child work is a consequence of a more reflective and subtle reaction to the millions of children in the world who must work.

Attempts have been made to distinguish between child labour and the work of children, which implies the banning only of the most dangerous and exploitative labour, while recognising that many families in the South depend for survival upon the income provided by their children. As long as the working day is limited and integrated with adequate time for education and play, the present wisdom goes, then the labour of children can also be a positive experience. It gives them a sense of involvement in the work of society, it raises their stature in the help they provide to their parents, it improves their skills for the labour of adults which they will one day perform.

There are objections to this over-neat formulation that distinguishes child work from child labour. In the same way, the demarcation between child-worker and child-student is also simplistic. Therese Blanchet says, 'The debate has been conducted over the heads of children, who are remote from policy-makers, theorists and researchers. The idea that work is bad, school is good, may also be contested.' Whether 'the learning entailed in labour' is better than 'the mechanical learning taking place in school' may be debated. Blanchet shows that under the labour/work distinction, the boys working in the *bidi* factories she studied in Kustia would qualify as child labour, since they worked long hours in a factory, whereas their sisters who roll the cigarette papers at home, 'in their spare time', for pay of only one-third of that of the boys, would be seen as child workers; and the injustice and gender exploitation of it would be lost in these arbitrary categories.

The discussion has developed further. It is now considered essential that children themselves should be consulted on their feelings about and attitudes towards work. Not surprisingly, it has been found that most children feel proud of the help they give their family, although many would rather be at school full time. What they want is to be defended against exploitation, violence and bullying, to be recognised as adult workers are, to be permitted to organise (as adults are sometimes allowed to do) and to be valued for the contribution they make, both to their family and to the wider society.

There is, perhaps, a cautionary word to be said about 'child participation', 'consulting child workers', 'letting the children speak for themselves'. For this relatively new faith in the wisdom of children is also a culturally determined and Western concept. More than this; it is also of a piece with all the other picturesque language emerging from government agencies, NGOs, international institutions alike, about the virtues of 'participation', 'people-centred development', 'empowerment', 'poverty alleviation', 'gender balance', 'income generation', and all the other popular ideas now offered to the poor of the earth. It is no accident that such concepts should be widely accepted at the very moment when those who control the macro-economy are more insistent that 'there is no alternative', that 'market reforms', 'liberalisation', privatisations, cuts in government spending, devaluation, elimination of budget deficits, may no longer contested as the only way forward.

That these stern realities strike at every turn against the more generous concepts of empowerment and poverty alleviation goes without saying; and given the structures of the global economy and its unappealable necessities, there is little doubt which of these will win, when, as is obvious, they must clash in the places where people have to make their own accommodation with a single global system. This is not to devalue the good intentions embodied in 'participation' and all the rest of it; but it is of not lesser importance to understand the context that may well undermine, if not cancel, them.

Implicit in all these discussions are other deeper assumptions. One of these is that even though child labour may remain essential for the time being, the developmental model – which is nothing other than a replication of the Western way – will eventually, by the creation of more wealth, sooner or later raise the living standards of every country on earth in such a way that the institutionalised employment of children will come to appear to the countries of the South much as it does in the West today – an archaism, a remainder of a more brutal and insensitive age, an unfortunate necessity which attends a lower stage of development than that which the West has now reached. Leaving aside the large number of child workers in the West (especially among immigrant communities and the children of poor families), this, too, begs many questions.

Not the least of these is whether all the countries in the world can attain levels of affluence that the West has reached. This once more reopens the story of *how* the West became rich: the territories it was able to annex, the lands, riches and resources of others, which it could call upon in order to bring certain basic amenities to its own people. The story of the extinction of indigenous peoples in North, Central and South America, Australasia, the history of conquest and control, cannot be separated from the model which is now universalised and of which no significant critique is thought possible. Yet no discussion of the future of the poor and their children should ignore the implications of the apparently unproblematic spread of the wealth-creating miracle to the whole world, without asking what underlay the success of the original model.

Indeed, present-day examples are not wanting in examples of attempts to colonise the lands and resources of others in order to feed economic growth and expansion. One of the most obvious examples has been Indonesia, which faithfully continued the policies of its former imperial masters by annexing East Timor and Irian Jaya and, by its policy of transmigration, decanting the 'surplus' people of Java into the territories of others, just as Britain could send indentured servants, convicts and others to North America, South Africa and Australia. Similarly, Brazil has sought to enlarge its area of cultivation by its violent incursions into the Amazon, with the consequences there for the remaining indigenous peoples with which the world is familiar. But such brutal options are simply not open to the majority of the world's countries in their efforts to make their economies expand and grow.

Even so, the logic of perpetual expansion compels even poor countries to mimic the colonial economic model preached to them. Bangladesh sought to settle Bengalis in the areas of the country traditionally occupied by tribal people, notably in north Bengal and in the Chittagong Hill Tracts. The military governments of Zia-ur-Rahman and General Ershad offered inducements to 'settlers' in these areas. The absence of traditional land records among the indigenous peoples made it easier for incomers to stake a claim to areas which were regarded as belonging to no one. This led to twenty years of insurgency in the Hill Tracts and, in the process, destroyed the environment and culture of the Chakma people and of the other ten tribal groups which had coexisted with them in the hills. The trees have been felled, exotic species planted, the culture of the original inhabitants broken, its fragile belief system destroyed with the ruined environment. In such places, the nature of globalisation can be seen for what it is and, whatever benefits flow from it, it is impossible to avoid the consequences for its victims.

It is precisely these difficult – and often unanswerable – questions which lead to a certain evasiveness in the recommendations of many NGOs and international institutions alike. The ILO representative in Bangladesh, for instance, accepts the gradualist position in the removal of child labour. Dr Wahid ur Rahman says:

Many NGOs and professional organisations initially took the stance 'Eliminate child labour'. But most realise it cannot be done instantly. The industrialised countries did not develop into a situation overnight where there was no child labour. It was a gradual process, a changing culture made it unacceptable, and an economic situation where there were good alternatives for children. Skill training, economic growth and development are all connected factors which together contribute to the reduction of child labour. Most now accept this, and only a few hardliners still insist that it must be stopped swiftly. We now speak of intolerable labour as against work. The new ILO Convention on Child Labour takes a more realistic line on what can and cannot be done. We speak, first of all, of child work, which is distinct from child labour, which is itself distinct from intolerable child labour.

If a child has a normal childhood, attends school, is working maybe five to seven hours a week, ten hours a week, even more, that is work, say making cakes in the family and selling them, that is okay. Child labour should be eliminated over time, a timetable of twenty years, reflecting a change in the educational system and the development of pre-vocational skills. Intolerable child labour is the immediate focus of concern, and involves the ending of children employed in dangerous and hazardous labour.

A new acceptance of child work under the banner of 'realism' and a reduction in the urgency about eliminating child labour mark a desire to remove the focus from a globally unjust system which fails to provide hundreds of millions of adults with a livelihood adequate for the nourishment, education, welfare and care of their children. Why this should be so may be traced to the growing gulf between rich and poor in the global economy. In 1820, the ratio in living standards between the very richest and poorest countries in the world was about three to one. By 1913, this had risen to eleven to one. By 1950, it was thirty-five to one. In the late 1990s, the richest countries of the G-7 were more than seventy times richer than the poorest countries of the Sahel (see Bloom and Sachs 1998).

No wonder the talk is now of gradualism, of the long-term, of distant goals, of aspirations and hopes for wealth creation. Hence the justification for 'tolerable' child work. After all, if we consult the children, they want to work, they feel proud of furthering the welfare of their family. A growing opacity clouds the arguments, not because these are complex, but because relationships of injustice must not be disturbed from their secular slumber. Of course, it may well be that children do need to have a social function to make a contribution to the work of society; but that is not necessarily the reason why this tolerance is now being rediscovered by the ILO, UNICEF, Western governments and financial institutions. These agencies have revived older attitudes and sentiments about whether work is a proper occupation for children. This is sometimes expressed in an extreme form – those who believe that labour is good for discipline, instruction and obedience in children, and those who believe that childhood should be a labour-free zone, reserved for play and learning, which are the indispensable prerequisites for the making of a balanced adult. But increasingly, acceptance of child work takes the shape of an argument in favour of cultural diversity. It is expected that children in South Asia should work; and who are we to interfere with the wisdom of such ancient custom and practice? Again, there may well be truth in this, but the reasons why such wisdom is now being heard once more should also be questioned.

There can be little doubt that many employed children are acquiring skills which not only help their impoverished families now, but will also stand them in useful stead for the rest of their lives. But many such activities cry out for some kind of regulation, for greater protection of the children from dangerous machinery, for control of the number of hours worked, for some superintendence of the power of owners over them. It is not that such activities take place in secret places remote from the eyes of officials. It is all on the streets, open and obvious.

Near Kamalapur in Dhaka, along the high walls that protect the railway line from encroachment by the homeless, children are working in groups on the pavement, in full glare of the fierce sunlight. Most are breaking bricks – this must be the commonest employment for children in Dhaka; indeed, its very prevalence gives certain areas of the city the aspect of a vast correctional institution, or a hellish afterlife of punitive activity for misdeeds on earth too terrible to be recorded.

Sibhu is eight. Barefoot, he wears a pair of dusty shorts and a faded shirt without buttons. He is earning 20 taka a day, working five hours and attending a government school in the afternoon. He has four brothers and one sister. His father is a peon in a government office, his mother a domestic servant; but although both parents work, their combined income does not provide enough to feed the family adequately.

Aklima is six. She wears a necklace of pearly beads over a floral dress discoloured by sunlight and made dingy by brick dust. She breaks bricks with her mother and two older sisters. Together, the family earns between 50 and 60 taka a day. They do not go to school. Aklima says she has 'always' done this work; her short memory cannot remember a time when she did not.

Opposite this wall, there is a long row of small workshops. These are sizeable concrete shells, oily, dusty, festooned with ancient cobwebs. Each of these employs five or six young men and boys. From each one, a tapping and beating of metal, pools of iridescence from spilt oil, a clattering and banging of panels, engines and fenders.

Hanif is learning to repair baby-taxis (that is, the two-stroke engine Bajaj three-wheelers imported from India). Hanif is twelve and earns 20 to 30 taka a day. He does not go to school. He has three sisters, one of whom is married, while another works in a garment factory. His father is a construction worker. Hanif has already been working for five years, for the first three of which he earned nothing. He has learned by watching the older boys and has become skilled through practice. The family is landless and came from Barisal. They live in a slum behind the main road.

One of the questions that arises in Dhaka, as it did in nineteenth- and even eighteenth-century England, is over the nature of the training 'apprentices' actually receive. It is in the interests of employers, under the pretext of 'training' to restrict their activities to a few simple tasks which do not amount to a coherent skill. In this way, their period of subordination may be prolonged, and they provide cheap – or free – labour to the workshop owner, sometimes for three, five or seven years. Dorothy George in *London Life in the Eighteenth Century* (1925) wrote of the apprenticeship system:

> There were frequent complaints that the master, instead of teaching a boy his trade, would keep him employed on some labourer's routine work or as a household drudge. There were often times when the small master or the journey-man taking job-work had no job to do, and either hired out the apprentice to someone else or allowed him to roam around the streets to pick up a living as best he could. The poorer, the younger and more friendless the child, the greater, of course, were the dangers and miseries of apprenticeship.

Jasim is nine. He operates a lathe, making parts to restore broken and crashed cars. Pieces of old metal are used to make new panels. The rusty shell of an old Ambassador car stands on the concrete. Jasim has been working for only a few months, and so far has earned nothing. He has two brothers and five sisters, two of whom are married. One brother is in the Gulf. His mother and father stay in the village in Faridpur, where they own six decimals of land. Jasim has come to Dhaka to live with his uncle who is the owner of the garage. He has never been to school, but has faith in the relative who will provide him with a trade. The lathe which Jasim is learning to operate is an unwieldy object, encased in blue metal, imported from India. It requires all his child's strength to put the sheet of metal in place, tighten it with the help of a metal wheel, and then bore it with a drill.

Noyon is Jasim's uncle. He explains he is not really the owner, but is managing the garage for his brother who is working in Qatar. He judges it will take Jasim three or four years to learn the job thoroughly, but he will start earning 500 taka a month after one year. Noyon himself started work through the same unpaid apprenticeship system. He knows of unscrupulous employers who teach the boys nothing of value. But he is Jasim's blood relative, so this couldn't happen. They bring their own food each day and eat as they work. Jasim and his uncle live in a sublet room for which they pay 1,000 taka a month. The adult workers are well paid – 3,500 taka a month (about US$ 70). Noyon says that the cost of running the garage is about 15,000 a month. The profit is 10,000 a month, from which he takes his salary, while the balance goes to his brother.

Chapter Nine

Contemporary ambiguities about the value of child work also have historical roots in the West. Writing of the industrial revolution, Asa Briggs (1959) says:

> The employment of women and children was not a new departure in industry. Both had been compelled to work hard in the countryside and in domestic industry. Children's labour had been regarded as morally useful as well as economically necessary, and even the benevolent Rev. David Davies recommended that no persons should be allowed parish relief on account of any child 'above six years of age who shall not be able to knit'. What the factories did was to spotlight the problem in an age when the number of children in the population was increasing rapidly.

Many observers of the countries of the South make the point that the cities are teeming with children. The attack by Western countries on child labour is, among other things, a reflection of their own ageing populations. Not only has this made youth and children objects of great veneration in the countries where they are a low proportion of the population, but the numbers of children in the Third World also makes the West uneasy, in terms of strength and future fertility: the population 'problem' that is to such a degree a projection of, the obverse of, Western over-consumption.

Whether we look at pre-industrial Europe or contemporary South Asia, it becomes clear that the work of children was widely perceived as both essential for the family and salutary for the development of children. William Howitt, in *The Rural Life of England* (1838), wrote that village boys were set to work as soon as they were old enough to look after themselves. Often, their first job would be watching a gate

> that stands at the end of the lane or the common to stop cattle from straying, and there through long solitary days they pick up a few halfpennies by opening it for travellers. They are sent to scare birds from corn just sown or ripening ... They help to

glean, to gather potatoes, to pop beans into holes in dibbling time, to pick hops, to gather up apples for the cider mill, to gather mushrooms or blackberries for market, to herd flocks of geese or young turkeys, or lambs at weaning time; they even help to drive sheep to market or to the wash at shearing time... (cited in Pamela Horn 1976:61)

By far the majority of children working in the South today are performing the kind of labour which their forebears did in Britain in the late eighteenth and early nineteenth centuries. The oral testimonies that have survived speak of uncanny convergences, which easily surmount differences of culture, time and climate. *The Hungry Forties* was published by T. Fisher Unwin as an anti-protectionist tract in 1904, a compilation of memories of people who remembered the 1840s. However tendentious its message, the pages are crammed with the authenticity of direct experience; and the evocation of British country childhoods in the first half of the nineteenth century foreshadows the life of millions of children in present-day South Asia.

I'd very little schooling. many and many a pail of water I've carried up the hill for a penny, aye, even a halfpenny ... I worked in the meadow for threepence a day, spreading the manure and picking the stones; and afterwards I got swede-cutting ... I minded the sheep sometimes and got threepence or fourpence a day for it.

I used to go cow-minding and bird-minding at threepence a day, or 1s. 6d. a week.

One man in his hundredth year in 1904 wrote:

I was born in Maresworth, Hertfordshire, in 1804, and worked early. I worked as a plough-boy, with my mother's boots tied to my feet with string. My first engagement was with a farmer, who, in return for my labour, gave me free food and no wages. When I was too ragged to be decent, my master applied to the parish for clothes for me. We used to wear sheepskin breeches, and when we got them wet through, we lay on them at night to dry them for the morning.

This was the early work experience of a boy from Khulna, Bangladesh, now working as a rickshaw driver in Dhaka.

I worked when I was very small. I had to look after the animals, see they didn't stray. I was out all day. It was very lonely. I took some rice in a bag. In the summer it was good, you take fruit from the trees, especially at mango time. But in the rain, sometimes I walked with my knees in water, and there were worms and infections in the foot. Snakes were the most dangerous thing, and you had to be careful. If you got a snake bite and you were a long way from home, you knew it could kill you.

A man from Leicestershire stated:

> I was born in 1836. I was sent into the fields to scare crows, and when I had done a full week, seven days, I had one shilling. I was such a small boy, my father carried me on his back to work … I was the oldest of seven children; and when I was old enough I crept into the wood by the light of the moon, and brought out once five pheasants to help to keep my father, mother, brothers and sisters from starving … The farmer was the jackal of a landlord.

Hamida is now thirty-five. She came from Bogra in the poor north of Bangladesh to work as a maidservant in Gulshan in Dhaka. She was the eldest of six children born to a landless labourer. From the age of seven it was her duty to look after her siblings, to cook and prepare food, and look after the house, a hut of wood and bamboo. The other children all began work at six or seven in other people's fields, looking after the rice crop, weeding and, later, transplanting. When Hamida first came to Dhaka, she was very frightened. 'I didn't know if I would be beaten or starved or scolded.' She was married at thirteen but it was understood that her husband would not touch her for two years. She was told, 'This is your husband.' 'I was a child. I knew nothing else. I had never left my village.' When she speaks of the journey by bus, the clenched terror sitting in the seat, her sense of apprehension was almost certainly that of little girls who sat in the carrier's cart, or the train taking them to London from the Midlands countryside in the 1850s. Hamida has now bought a little land. When she went home for the first time, she took the money with her in a bag. 'My mother almost fainted. I said to her, "Touch it." We never saw money. My youngest sister went to Dhaka to work in a garment factory; she was retrenched at fifteen and went home. She is now looking after my three children, together with our mother and grandmother.' Hamida always hated field work. Because she was the eldest in the family, it was her job to look after the five younger children. 'I was the only one never to do farming. But there was no end to the work I did each day; when I came to Dhaka it was no hardship for me, because it was only what I had done all my life.

The only difference was that I was doing for strangers what I had always done for my family.'

A man born in Sussex said:

All the schooling that I ever had was at a dame school, Mrs Baker was her name; she lived up the common. She couldn't write her own name, but she could read a bit and she taught us to read. When I was rising nine, my father said I'd had scholarship enough, and sent me to work at the sawing, and I haven't been out of work more than a day since.

A few kilometres outside of Dhaka, beyond the brickfields. The first settlements that are not city, even though their rural appearance is an illusion – behind the clumps of trees there are factories, a Bata shoe factory, garment factories. Some children are working in a field of *dhania*, which is not yet ready for harvesting. They are collecting fodder, taking care to cut only the coarse grass and making sure they do not trample the *dhania*. Two boys are working side by side; the rasp of the sickle can be heard in the still afternoon as it cuts the tough grass. Two boys are working with their father who has four children and three cows, but no land. This spot where they are cutting fodder belongs to a man in Dhaka. They do not have his permission to cut grass. If the owner discovers that they have been here, he will be angry. Pondab, the older boy, goes to school in the morning and then helps to collect food for the cows. This is a daily chore. The father says that his main occupation is shoe-mending. He has a box of tools and polish, and goes to Dhaka almost every day, where he can earn between 40 and 80 taka a day.

A second child, Polash, who is seven, works alongside them. He does not go to school, but collects grass for his family's two cows on their one-third acre plot. Polash has close-cut hair, and he stoops, barefoot, in the grass. 'Education', he says, 'I can get later. The cows have to be fed today.' He smiles at his child's ancient knowledge of daily necessity and returns to his work.

I remember distinctly before I was eight years old having to spend the bitter cold winter days in a large field scaring rooks, and as fast as my little legs could drag over the heavy clay field to one side, the rooks were on the other side; and many a bitter tear I shed over my failure to scare them ... I have a very distinct recollection of dumplings made of barley meal, and it was with some difficulty I got my teeth through them. Then we had some potatoes, and sometimes found a swede in the road, having fallen off the farm cart. That was a treat indeed! Meat? Oh no! A simple herring between five us constituted our Sunday dinner, and the tail, I remember, was my share as a rule. Ah! And then we had tea – sugar we hardly knew the taste of. This tea was such a lovely brown colour; and one day, being rather curious, I thought I would find out what it was made of, and looking into the teapot, I found some burnt crusts of bread. This was our lot of semi-starvation and slavery.

Other children in the field outside Dhaka who go to school say they must always be ready to do whatever work is required as soon as they come home. They fetch water, help prepare the food, look after younger children. Their father sells chappals which he buys directly from the nearby factory. They say that finding fodder in the dry season is a long and difficult job. The alternative would be to buy in the market – 22 taka for two and a half kilos of hay. To feed cows is as expensive as feeding people. The cattle do not go hungry, but people sometimes do – on occasion, their meal is rice, chillies and salt.

'For years', a Norfolk man testified, 'I never knew the colour of money. I worked in the mill, and was allowed a certain amount of flour each month in lieu of wages, and even then I did not get enough flour to meet the wants of my hungry family.' Another, 'born in the year 1830, in a country village in Essex, put to work at the age of nine years – many of the children of the village at work before that – for the pay of 1s a week. No school in the village, only kept by an old dame ... numbers of children had no learning but what they got there; lads from twelve to fifteen years of age learning to spell small words, such as "I cannot see God, but God can see me" ... the first lesson after learning the alphabet.'

A man born in Plymouth in 1826 says:

At six years of age I was sent into the fields with the apprentice boys and girls of about the same age, to keep pigs, clean turnips, drive oxen and horses at plough, and various other field work ... The delicate children were soon killed, in fact one was knocked over the stones by my great-aunt and her neck broken, and no notice was taken of it. She used to lash us with her riding whip for the least thing, and when my uncle came in, she would not let him rest until he had thrashed us as well.

Anwar is now twenty. He is from Faridpur and is the oldest boy of five brothers and four sisters. His father is a landless labourer, now too old to work. Anwar never went to school, but began work at eight as helper in the fields, weeding, planting and cutting rice. For this he was paid with a little rice if working alone, nothing if he was helping his father. Later, he worked as helper to the *ferrywala*, the boatman who carries passengers across the river. For this he was paid 15 to 20 taka a day.

> My father's native place was Honley, about seven miles from Huddersfield. His parents were poor working people – so much so that they had to get rid of their children as best they could; so my father was an apprentice to a farmer – he got his food but no wages at a village, Crosland Hill, his master finding him what clothing he thought useful, while he was of age. After his apprenticeship he went to work in the stone quarries. In due time, he got married, and there was a family of three children. I was the second and had two sisters. Poor mother died when I was between two and three. My eldest sister went to work in a factory very early. I soon had to follow, I think about nine years of age. What with hunger and hard usage I bitterly got it burned into me – I believe it will stay while life shall last. We had to be up at five in the morning to get to the factory, ready to begin work at six, then work while eight, when we stopped half an hour for breakfast, then work to twelve noon; for dinner we had one hour, then work while four. We then had half an hour for tee, and tee if anything was left, then commenced work again on to eight thirty. If any time during the day had been lost, we had to work while nine o'clock, and so on every night till it was made up. Then we went to what was called home. Many times I have been asleep when I have taken my last spoonful of porige – not even washed, we were so overworked and underfed. I used to curs the road we walked on ... We had not always the kindest of masters. I remember my master's strap, five or six feet long, about three-quarters of an inch broad, and quarter-inch thick. We could not mistake its lessons; for he got hould of it in the middle and it would be a rare thing if we did not get two cuts at one stroke. (Unwin 1904)

Minara and Rumi both started work in garment factories when they were eight. They are now sixteen. Minara is now earning 100 taka a month (about $US 35) at Oasis Garments, which she joined as an operative three months ago. She has not yet been paid any wages. At Eid she received 200 taka (four dollars). Minara went to a Koranic school and can write her name. She was married a year ago. Her husband is a construction worker and earns 150 taka a day. Minara still lives with her mother and father, because she and her husband have no place of their own.

Rumi also started work 'at eight or nine', at Patriot Garments, and was paid 150 taka a day (three dollars). She was very frightened by the factory, overwhelmed by the noise and crowding of the workplace. The family had left their village because Rumi's father had become involved in a court case – a dispute over a piece of land that led to a fight and injuries. Rumi is now working with Minara at Oasis Garments. She is yet to receive any pay. Rumi was married at six, but did not go to her husband until she was fifteen. She is still living with her family in Sipahibag. Her husband works in the same factory. Rumi can write her own name and read 'a little'. Rumi wakes up at six thirty, leaves the house at seven to be at work by eight. If the day finishes by eight she will reach home by nine. Sometimes it is ten o'clock before they finish, and she arrives home at eleven.

Minara's father comes to meet her if they are late. The young women walk on the main road if they are late, moving in groups so there is less danger of being molested. The factory owner treats them roughly. There is no place to eat, no place for prayers, insufficient toilets for the 200 workers. Water is available 'sometimes'. When there is an order to be completed, they may work through the night until five o'clock in the morning.

The fields of Bangladesh are crowded with children. Barisal, the watery South of the country. A boy of about eight stands with a buffalo in a pool, wiping the dirt from the creature. He splashes water over its back, lifts the tail to wipe the encrusted grey shit from its backside. Two little girls of about seven and nine, barefoot and carrying a stick, are looking after goats. A boy of perhaps seven is lifting the retted jute from a canal beside the road, detaching it from its stalk and laying it in neat bundles on the grass. Two Adivasi children, wearing conical bamboo hats, are minding a herd of black pigs. A child stands up to her waist in the river catching shrimps with a net. A boy of eleven collects the fares on the small ferry that crosses from Barisal to Khulna. Children are selling bananas, papaya, drinks, sweets at the ferry terminal. A girl of about nine is almost invisible beneath her headload of fodder; her hips oscillate with a graceful tension as she walks, almost overwhelmed by the weight. A boy of about five carries a headload of firewood, broken pieces of tree in a bamboo basket. No one in the countryside is not working or engaged upon some purposeful activity.

At least the landscapes, both of early nineteenth-century Britain and of present-day Bangladesh, have the familiarity of known and loved places: they are peopled with kinsfolk and family, neighbours and fellow-labourers in the field. Whenever my mother recited the stories of our relatives spread out in the countryside of Northamptonshire, there was a great web of kinship, a living tissue of humanity which sustained and bore up its members: this is how it must be for the people of Barisal or Rajshahi or any other country place in Bangladesh. What changes the comforting fabric of rural life is its growing impoverishment, the ending of self-reliance, the growing dependency upon goods, services and money from elsewhere.

This alters the perception of the reassuring environment: the ground shifts, it no longer gives comfort, but anxiety; it is no longer a source of security, but hostile, insufficient to give sustenance to its people. The people do not change, but everything changes around them, so that the environment is still, and yet is not, as it was. It is incomprehensible, disorienting. People resist moving, try to remain in the ancestral lands, hoping that it will all become as it was before. But it doesn't. The earth itself becomes hungry for strange new forms of 'food', fertilisers and pesticides, so that it is no longer by the sweat and work of the labourer that it produces paddy, but through a money investment. The peasant sensibility strikes against industrial necessity like a sickle trying to dig through flint.

The people stolen for slavery were also placed in an ugly and alien caricature of village life; but for them, all custom and tradition had been effaced. That they managed to reconstitute something of the broken structures of kinship and belonging says much for the spirit of human beings in the least propitious circumstances imaginable. Walvin evokes the slave villages of the sugar islands.

The slave family lived in huts or cabins, 'negro houses', which varied enormously. On the bigger plantations, especially after 1750, planters grouped their slave families in huts huddled close to the fields they worked in. Earlier, in the predominantly male slave force, they had sometimes been housed in barracks, catering for six men or more. But on smaller farms, slaves slept and lived where they could find room; in lofts, in the tobacco houses and in a host of outbuildings.

From their earliest years, slaves' lives were shaped by the peculiarities of the slave system. Mothers cared for their children up to the age of ten; over the next four years children left their parents' home; some settled with other brothers and sisters, some were sold, while others moved in with other relatives. Slave women tended to marry in their late teens and then establish households of their own. The newborn was generally taken into the fields with the mother. When the child grew older he/she was left in the company of other small children in the slave yard, possibly under the supervision of an aged slave. And like rural children the world over, as they grew a little older they were expected to care for the smaller children, allowing the mothers to work uninterrupted in the fields. By the age of seven, the children were ready for their initiation in the local fields, learning the necessary skills and tricks of the trade from parents and white overseers. In the same way, the slave craftsmen passed on their trade to their growing progeny. It was at this point, when the young slaves had begun to work and had an economic potential of their own, they were removed from their family home and placed in what their owner thought was more appropriate labour.

The consequences for children who are 'free labour' in Bangladesh does not depart significantly from that of their distant slave kin in the Caribbean islands.

Apart from the most privileged classes, the work of children has historically been the norm; what is perhaps more aberrant is that childhood is now a place apart, abstracted from the work of society. That children should be without function is a novelty, and given the levels of concern in the West for the troubled lives of so many young people, it may be that their privileged social uselessness is a significant element in the cause of much disorder and discontent. Somewhere, between the overworked infants of Bangladesh and those indentured to consumerism in the West, there must be a proper balance of work and leisure; only, like treasure reputed to have been buried in some undiscoverable yet familiar spot, no one seems able to say where it lies.

Chapter Ten

Despite cultural, religious and climatic differences, the rural poor are always poor in the same way. Therese Blanchet, in her book *Lost Innocence, Stolen Childhoods*, describes how the present-day Western sensibility fails to understand the cultural differences between children here and children in Bangladesh. This is undoubtedly true. Yet if we look at some of the historical evidence, we see a far greater convergence between pre-industrial Britain and other predominantly rural societies today: Blanchet's observation is as much a comment on our severance from our own past as it is a comment on cross-cultural misunderstanding.

Blanchet says:

> I shall like to comment on a widely held view expressed on studies of Western childhoods. The concept of otherness to qualify the relationship between adults and children does not seem to apply in Bangladesh as in the West. Perhaps this is because there are other social dividers which are more significant than those between adults and children. A child bride, a child prostitute, or a child domestic servant may be constructed as 'the other', but this otherwise is not primarily an attribute which centres on their being children.

Philippe Aries, in his *Centuries of Childhood*, showed that children, when depicted in the iconography of the Middle Ages, appeared as miniature adults. He describes the emergence of childhood as a recognition that the child needs to be specially prepared for life 'as a kind of quarantine before he was allowed to join the adults'. Much of what Blanchet observes in Bangladesh would have been familiar to pre-industrial Europe. She states, 'There is no Bengali word to describe life from birth to the age of eighteen, like minor. "Shishu" as a child means one who needs milk and it implies dependence. A child who can fend for himself or herself is not a shishu.' This means that the idea of Rights of the Child commands ready assent in Bangladesh, because it refers only to helpless children. Those who are independent, working, married, etc., are not covered by the term. This makes the Rights of the Child problematic in Bangladesh. She says that 'to understand' is a key notion which sums up what it means to grow up. Often, this means knowing what is morally good; when children are capable of understanding and internalising the moral norms of the society.

That something similar was at work in Britain may be seen from attitudes towards child criminals in the early nineteenth century. In *The Hanging Tree*, V.A.C. Gatrell describes how, in 1831, an illiterate and pauperised country lad was sentenced to death for murder. 'Early on the Monday morning following, he was taken out of Maidstone prison to a scaffold. Four or five thousand people assembled to watch ... Afterwards his body was dissected by surgeons, It was all common enough for the times, except that John Bell was aged 14' (p. 1). In a study in 1995, *A Critical View of the Judicial Institution in Relation to the Rights of the Child*, Rohfitsch discusses juvenile crime in South Asia. He observes, 'The age of the offender is rarely a consideration and the nature of the offence is the paramount factor. Children are not regarded as being worth treated differently.' Gatrell records the revulsion of the prison ordinary (minister of religion) in the early nineteenth century at Newgate prison, who 'was indignant when he found boys of eight and ten in the prison for stealing pots, astonished when he found two female children of very tender years under sentence of death, in the women's condemned cell, and another one for trial – and these facts are recorded in the year 1824 in the Metropolis of England' (Gatrell 1994:385).

In relation to child labour, Therese Blanchet refers to children being 'given' by their parents to the factory. They are not consulted.

> They go willingly, even if they hate the work. They see a responsibility in earning. They owe it to those who gave them birth and brought them up. They also feel proud of their contribution to the family livelihood. Some parents believe that their duty to provide for their sons ceases when these reach the age of eleven or so. It is assumed that 'a good son' remits his entire earnings to his parents until he is fifteen or so, then he may keep some pocket money.

All this comes as no revelation to those who chronicled the lives of child workers in nineteenth-century Britain. E.P. Thompson, in *The Making of the English Working Class*, says that childhood was an intrinsic part of the agricultural and industrial economy before 1780, and remained so until rescued by the school. 'The most prevalent form of child labour was in the home or within the family economy ... The child's earnings were an essential component of the family wage. It is perfectly true that the parents not only needed their children's earnings, but expected them to work.'

Even within living memory, child workers remember that they used to 'tip up my whole wage packet to my mother. As we got older, we might be allowed a penny in the shilling for pocket money, but you didn't think you had a right to your earnings' (mill workers, Blackburn, Lancashire, in conversation with J. Seabrook, 1969).

'The idea of indebtedness is instilled in children; the rights of parents or guardians over the wealth children produce is absolute.' This could have been a description of the manufacturing districts of in Britain in the early nineteenth century. Yet it is Blanchet talking about Bangladesh in the 1990s.

We make assumptions in our view of social progress that render our own recent past unintelligible to us. This forgetting of our own experience has a more sombre consequence: it permits us to look with uncomprehending eyes upon those who endure the same violence which was commonplace not long ago in our own culture. This creates barriers to an imaginative understanding of their sufferings, useful doubtless to those who would have us tolerate these abuses, but unhelpful to the relief of the burdens they must endure.

My mother and her ten siblings all began work in boot and shoe factories between 1893 and 1919. The oldest began work when she was eleven, but by the time my mother, the youngest, started in 1919, she was fourteen. The children began as 'knot-tiers', that is, when the machine had finished the stitching of the shoes, they tied knots with their fingers in the threads. They sat on stools close to the machinists. When I first saw the child thread-cutters under the sewing machines in the factories of Dhaka, I was shaken by a flash of instantaneous recognition. It was only then that I really understood how it must have felt for my aunts to have been tethered to the only form of employment known in our town at that time.

Such strong ties of kinship and continuity across time and culture, a suffocating sense of something familiar: not a vague apprehension of something known, but a precise incarnation of forms of suffering that affected the bodies and sensibilities of children in Northampton in the 1890s and continue to affect those of children in Dhaka in the 1990s. Such a shocking paradox: when I go to Northampton now there is no trace, no memory of that experience, but to go to Dhaka is to go home. Who would have imagined that this journey to Bangladesh would be a pilgrimage to the same flesh and blood, inseparable from that of the long dead, bearing the same marks and scarrings, nourishing the same frail hopes and poor dreams of a better life.

There are even older echoes in our own memory, too. My grandmother's sisters left their village in Northamptonshire at the age of eleven in the mid-nineteenth century to work in service in

London. One of them went to a place in Eaton Square, then, as now, a highly fashionable area, on the recommendation, no doubt, of local gentry in her home place. When they spoke of their terror, as they sat, with a case on the back of the carrier's cart that would take them to the railway station in a town they had never seen, which would in turn convey them to London, the same tremor may be heard in the voice of young women transported to Dhaka to work in service or, indeed, in the garment factories.

A young woman I spoke with in Dhaka early in 1999 had begun at the age of eight. She described her first day in the factory. 'Everybody was so busy, I was frightened of the machinery, frightened that everybody knew what to do and I didn't understand what was happening. I was afraid they would beat me.' Children's fears that could not spoken, since travel to an unfamiliar place to perform strange tasks in an unknown environment represented their contribution to family survival: even their absence from loved flesh and blood was an indirect help to the family which they had left in such sorrow, for apart from the question of whether or not they were able to send money home, this also meant one mouth less to feed.

Even in Britain, this is not entirely experience from time past. In 1998 I interviewed a woman of eighty-eight, the daughter of a farm labourer, who had left her village in Kent to go into service in the house of a doctor in North London. She had a small room at the top of the house, entirely without heating, and worked from early morning until late at night, with one half-day off on alternate Sundays. The children in her care were only four or five, but they were imperious and made her feel her dependency and subordination.

There is nothing new in the humiliations and indignities heaped upon the children of the world. Perhaps we have erased from our collective experience all memories of that pain which our children knew, in order not to be reminded of it in its continuing existence in the world. Is this, perhaps, because the ending of the suffering of our own people has been directly dependent upon its transfer to them, the thin shoulders of the little girls of Dhaka, the cattle-minders of India, the child factory workers of Indonesia? A collusive forgetfulness obliterates our kinship with them; to which another colour, a different climate, a separate religion and a foreign language also eagerly invite us.

Chapter Eleven

Barisal, in the south of Bangladesh, is a place from which people migrate to all over the country. Barisal Town is the main urban area in one of the beautiful parts of the country. Close to the Bay of Bengal, water threatens at every moment to overwhelm the land, with its fragile embankments, its *chors*, or islands, that appear when the floods recede, its canals and rivers. The rice is beginning to ripen – at a distance it has taken on the yellowish sheen that gave the name of a mythic Golden Bengal. This area used to produce some of the most fragrant rice in the country, until overtaken by the Green Revolution. It is still covered with trees, and there must be at least thirty distinctive shades of green in the post-monsoon landscape – from the acid-green of algae blooming on still ponds, to the silver-green of the water hyacinths, the green-black of the mango leaves, the green-gold of the ripening crops.

But Barisal Town is a melancholy place, a fitting capital for a poor region, which is now a major exporter of its wealth, in the form of its people, to Dhaka, to the Gulf, to America. It rained for the whole week I was there. Puddles of olive-coloured water in potholed streets; buildings mildewed by long monsoons; ochre-coloured houses of former zamindars from colonial times, mouldering villas from the time when this was East Pakistan, with fretted brickwork, speckled and stained with lichen and moss. New cement buildings with tin roofs on which the rain crashes loud as gunshots; a blue-washed cinema hall, with lurid posters of plump crimson heroines and purple-faced villains.

There are square ponds all over town, overflowing in dark green wavelets blown by the humid wind; vendors under sodden half-collapsed umbrellas, mixing potions in small opaque glass bottles – remedies for rheumatism, fever and syphilis – selling rain-washed apples from India, repairing broken chappals, offering some meagre wares – locks, keys or cigarette lighters. Even the shops that stand open to the street are without customers, the dripping rain forming a curtain of silver beads in front of their display of watches, clocks or marriage jewellery. Medicines from Roche or Ciba-Geigy, manufactured in Karachi or Mumbai, in silver foil strips in white and orange boxes behind rusty metal and dusty glass.

Men walk barefoot in the mud, *lunghi* held in a graceful fold between thumb and forefinger so that it does not get splashed. The ubiquitous cycle-rickshaws are in constant movement – the relative absence of motorised traffic allows you to hear the vibrating chassis and the rattle of wheels, while the ringing of bells makes a strange pre-industrial sound of human energy as the wheels buckle under rough stones or get stuck in deep mud; red plastic seats, dainty silver metal canopies look as if they are heading towards some celebratory fete. Many of the drivers are very young: a fourteen-year-old is driving instead of his father who is sick with TB; another says he is twelve, with barely adolescent limbs and muscles.

Little restaurants with muddy floors and footprints of naked feet in which eight- and ten-year-olds serve. Rubbish overflowing choked ditches; plastic bags bobbing on the stretches of water like black and pink bubbles, workshops exploding in a radiance of blue and rose-coloured sparks, a Roman candle of yellow in the rainy indigo dusk. Kerosene lamps penetrate the wet dark, while a urine drop of pale yellow illuminates a vegetable stall where the cauliflowers are rotting in the damp heat. The Christian cemetery with its grass and goats and leprous tombstones which have weathered the existence of those they commemorate into a distant memory only two or three years after their death.

Barisal speaks of a grudging insufficiency, a makeshift city, a service town with no industry and not much service either, a market town for an agriculture that has become an extractive industry. Indeed, it is a kind of refugee camp itself for evictees from the rural area who are learning the new seasons of money, the unreliable harvests of day labour on construction site or brickyard; displaced persons, the multitudes of the no longer self-reliant, compelled into the service of the abstraction of money – as though the feudal lords who formerly controlled their lives had become spirits, no longer visible, but present everywhere, their power hovering over the markets, impersonal, insistent. Barisal, so many of whose young girls have gone to the sweatshops of Dhaka, girls whose youth and energy have been plundered for centuries have now left so that these brief qualities might even more systematically be used up than by moneylenders and local elites. Barisal, a miserable town, a jumble of iron and cement and stone, eaten by rust and mildew; a place to which no one comes unless it is to take away something else from the injured and impoverished people, to rob them of more land, more energy, more labour.

Barisal still lives off its local resources: woodyards, sawmills, country boats carrying jute, wood, vegetables along the backwaters. Many of the little shops are poorly stocked – soap, toothpaste, plastic toys, drinks, little packets of detergent, sweets, matches. On the corner of one street a shop is doing a brisk trade in biscuits

– a great variety of crisp baked confectionery, snacks for poor working people, which come from a nearby bakery. This is a long low building, a kind of hangar, about forty metres by twelve, where thirty or forty men and boys are working, The floor is rough concrete, the roof tin; there is a wide aperture that serves as a door and windows allow the air to circulate. There are two huge metal containers, one filled with flour and sugar in equal quantity, a second to which water has been added to prepare the mixture for the biscuits. It is stirred by men holding big wooden spatulas.

When the mixture reaches the right consistency, it is taken out and spread in tins, where it is cut, first into long slices, and then into individual rectangles, squares or circles, according to the type of biscuit. Much of this work is done by children. Two boys dip a pointed wooden stick into an orange-coloured sugar mixture, and they paint a decorative dab at the centre of each biscuit before it goes into the oven. A series of trays of raw biscuits, all being decorated by hand. On benches beside these, further trays of biscuits waiting to be cooked: some covered with chopped nuts, others running with sticky syrup. There are flies everywhere, a humming swarm, black beads in sunlight. Some settle on the biscuits and remain captive in the sticky sweetness. They move in shifting clusters over the vats of sugar and flour, sometimes indistinguishable from the dried fruit.

The ovens are just inside the open door: an opening about half a metre by thirty centimetres. Men with flat metal shovels place the trays of biscuits into the scarlet mouth of the oven, and withdraw them after fifteen or twenty minutes. Each variety of biscuit made here is displayed in a glass jar at the far end of the bakery.

Rahul is twelve. He works from eight in the morning until five at night. He has been here for a year. One brother works as a construction *mistri*, another is studying. Rahul stayed at school until class five. Dammia, who is eleven, has been here for two years. One of her brothers is a carpenter, another is at school. They say it is children's work to make biscuits. They earn 35 taka a day (about 70 US cents). When they are older they will find better paid work, maybe as rickshaw drivers or perhaps in construction. Md. Julemia is twelve. He has just started work. For the first few weeks, he simply observes and carries out simple tasks. He will not expect to earn anything in this period. The atmosphere in the bakery is stifling. When the sun appears it throws bright lozenges of light on the floor; the bakery shimmers in the heat from the ovens, and physical objects are discomposed by the hot fumy air.

Children – mostly boys – are visibly employed everywhere. A boy of about eleven sits in a small store, minding sacks of rice of varying quality: the neck of each sack is rolled back to display some pale white, some creamy, some dull and stony – the price varies from 14 taka for the most inferior to 30 taka for the best. Boys are busy in small hotels and metal-welding units. Beyond, unplanned private houses, a pond surrounded by bright emerald grass, a jumble of construction, where some people have created small homesteads beneath a plantain and jackfruit tree. Some houses have been half-constructed and abandoned: evidence of the volatility of work in the Gulf – a sudden retrenchment, a contract unexpectedly cancelled: the brick pillars and rusting metal rods are invaded by creepers with bright blue flowers. A garment factory, an apartment block still under construction, a piece of waterlogged land invaded by water hyacinths, foraging hens and ducks. Some very poor hutments of bleached bamboo.

In an open shell of a workshop, two boys are working alone welding some window grilles. Shaiful is twelve. He left school – like many children – after class five. He works twelve hours a day. Jehangir, his companion, is fifteen. He is arc-welding, a rough piece of smoked glass shielding his eyes from sparks. He has been working for three or four years. His family live about forty kilometres away from Barisal, but he visits them only once every two or three months. He lives in the workshop and serves as unofficial security guard. Shaiful lives with his family nearby: he is the oldest of four brothers and his father is absent. His 30 taka a day help to feed his siblings.

Nearby, one of the ubiquitous small wholesalers of junk and recycled objects, where children bring their daily industrial gleanings. A metal hut, stacked with paper, glass, metal and plastic. The owner sits on the threshold, as the children bring their darned jute sacks. He gives 10 taka a kilo for plastic, 6 for metal other than iron, 7 for iron, one and a half taka per kilo of glass, 2 taka for unbroken bottles, 3 for a newspaper. The rates seem to be higher here than in the slums of Dhaka. About fifteen children cluster around the foul-smelling hut. Sairuddin is eleven, has close-cropped hair and wears broken chappals. He has just received 25 taka. His father is a flower seller. Sheikhfarid is ten. He expects about 20 taka for an eight-hour day. He has three brothers and two sisters. His mother is dead and his father is a rickshaw driver. The children talk excitedly about their work. No, it isn't boring, because sometimes you find something valuable, even money. Their eyes are alert, they are always watchful, they say, even when they are not working. No, it does not disgust them to work among rubbish. Sometimes, they put a black plastic bag on each foot, a bag on their head to protect them against the filth. Once, says one boy, I found some gold from wedding jewellery. But when they tip up their sacks, it is a heartbreaking little cascade of junk – a plastic bowl, an iron bar, a clock face, a rusty key, a broken glass bowl, some pieces of opaque white glass, rusty nails, some wire, a split plastic colander.

On the other side of the highway, some men are clearing water hyacinths from the clogged canal that runs alongside the main road. They stand up to their waist in water, ankles deep in mud. The hyacinths – brought here by Lady Hamilton, because they took her fancy in the Mediterranean; within ten years, they had colonised the whole country – are used for green manure. The smartest building in the vicinity is the security shelter for the supervision of a new site, where a new building is coming up; a concrete rotunda, with whitewashed brick wall enclosing the compound. The space is empty apart from four women who, with their children, are breaking bricks. A truckload of bricks will earn the families 900 taka between them – three days' work. The broken bricks will form the foundations of the building. They do not know what is to be constructed here. They work from morning until evening.

The work of the boys is to carry the bricks from where they have been dumped by the truck to the places where their mothers work. They wear a pad of compacted cloth on their head so that the bricks do not hurt them. One little boy of about ten wears a red T-shirt that says COCONUT, shorts and green plastic chappals. Another is barefoot, wearing only a *lunghi* that displays an untreated umbilical hernia. The women sit on the heap of rubble which is the end product of their labour, under umbrellas that have been mended with plastic. A huge flat pebble serves as anvil. They wear rough protective rubber gloves to save the back of their fingers from badly aimed hammer blows; tapping and breaking, a flyaway cloud of red dust changing the colour of their faces.

Zuleika's two boys are Rincon, thirteen, and Feroze, ten. Her husband is a day-labourer. They sold their land 'for income' and came to Barisal. They live in a small hut for which they pay 4,000 taka a month. Zuleika has one daughter, fourteen, who has gone to Dhaka to work in a garment factory. She worries about her, but she sends 200 taka monthly. Zuleika's boys are not paid. 'It is their duty to help us', she says; and her simple sentence expresses the self-explanatory relationship between parents and children in Bangladesh. They do not go to school. Rincon wants to operate a lathe machine when he grows up, Feroze says he wants 'a proper job'. Holdibaby's husband is a day labourer, cutting soil. She says the four women share the 900 taka, but as collective work for three days, it does not amount to very much. I observe that Zuleika has two children to help her where Holdibaby has only one. Does that make a difference to the share of the money? She shrugs, as if to say, what can a child do?

Children are regarded by poor families not as individuals, nor scarcely even as an extension of adults. They are part of them, the same flesh and blood. It is this – to Westerners – archaic relationship that accounts for many incomprehensible actions by parents in South Asia. Children barely have an independent existence; and in this we may see both the strength and weakness of families who make no clear differentiation between its component flesh and blood; or rather, they see it as gendered, in the sense that boys are more welcome than girls; for boys – and the families they in their turn will make – will sustain them in old age, whereas girls, who will go to their husbands' home, taking a dowry with them, will only enrich the families of others.

There is powerful instruction here in the power of kinship: the identity of the extended family merges, held fast, as it is, in bonds of duty and necessity. The lives of the children have the poignancy of those who have not yet learned that it is the destiny of individuals to pursue their separate trajectory through time. It is both monstrous and touching; for, like so many experiences in the South, it evokes our own past and also relativises the breaking of bonds which were, in our country also, no less strong than they are now in Bangladesh. It interrogates our own norms, and poses questions about the separations and ruptures that have occurred between our people. Perhaps the most disturbing element in this is that the disregard in Bangladesh for the individuality of children is a mirror image of our own excessive concern with individualism. It seems that human societies are destined to oscillate between extremes, neither of which brings satisfaction or fulfilment, but which are part and parcel of the underlying belief system of the context, the social and economic base from which they cannot be detached.

Chapter Twelve

Whether we start with the reality of daily life in South Asia, or with the images that come to us from nineteenth-century Britain, the parallels are equally striking.

No one chronicled the labour of mid-nineteenth-century London in greater detail than Henry Mayhew. The ingenuity with which the people he met, drifting, scheming or toiling through the streets of the city found a place for themselves within the urban economy, prefigures the efforts of the poor in the cities of Africa, South Asia and Latin America in our time.

Mayhew did not single out the labour of children for specific attention. But they are everywhere in his narrative, always present, as ubiquitous as the urchins of Dhaka, Mumbai or Manila. Mayhew spoke with many whose livelihoods were of the most fragile and undependable kind – the marginalised collectors of dog-dung (used in tanneries), hawkers of birds' nests, bone grubbers, cigar-end finders, crossing sweepers, rat catchers, sellers of second-hand clothing, sealing wax, quills, pencils, braces, sponges, cheap jewellery, thermometers. Children weave and dart among this throng of adults in pursuit of precarious livelihoods. A seller of sponges told him:

> I believe I'm twelve. I've been to school, but it's a long time since, and my mother was very ill then, and I was forced to go out in the streets to have a chance. I never was kept to school. I can't read; I forgot all about it. I'd rather now that I could read, but very likely I could soon learn if I could only spare time, but if I stay long in the house I feel sick. I can't say how long it's been since I began to sell, it's a good long time; one must do something. I could keep myself now, and do sometimes, but my father – I live with him (my mother's dead) is often laid up. I've sold strawberries, and cherries and gooseberries, and nuts and walnuts in the season. Sometimes 6d a day, sometimes 1s; sometimes a little more and sometimes nothing. Oh, I don't know what I shall be when I'm grown up. I shall take my chance like others.

The employments described by Mayhew evoke in detail the work of their counterparts in South Asia.

> It usually takes the bone-picker from seven to nine hours to go over his rounds, during which time he travels from twenty to thirty miles with a quarter to a half-hundredweight on his back. In the summer, he usually reaches home about eleven of the day, and in the winter about one or two. On his return home, he proceeds to sort the contents of his bag. He separates the rags from the bones, and these again from the old metal ... He divides the rags into various lots, according as they are white or coloured; and if he has picked up any pieces of canvas or sacking, he makes these also into a separate parcel. When he has finished the sorting he takes his various lots to the rag-shop or the marine-store dealer, and realizes upon them whatever they may be worth. For the white rags he gets from 2d to 3d a pound, according as they are cleaned or soiled. The white rags are very difficult to be found; they are mostly very dirty, and are therefore sold with the coloured ones at the rate of about 5lbs for 2d. The bones are usually sold with the coloured rags at one and the same price. For fragments of canvas or sacking the grubber gets about three farthings a pound; and old brass, copper and pewter about 4d, and old iron one farthing a pound, or six pounds for 1d. The bone grubber thinks he has done an excellent day's work if he can earn 8d; and some of them, especially the very old and the very young, do not earn more than 2d or 3d a day.

At Narayanganj, on the ragged rural fringe of Dhaka, we went to one of the wholesalers of *tokai*, waste, junk and recycled objects. It was about three in the afternoon. The hut was crammed with stacks of paper, jute sacks of glass, metal, plastic and so on. The owner, a man in his forties, sat inside the hut on a battered chair in a small space which just allowed the corrugated metal door to open and close. Around him, clustered at the door were about six boys, all trailing jute sacks, some of them darned for re-use. They were without shoes; they had the skin of the children of the poor – marked by the scars of cuts that had not healed properly, insect bites, remains of untreated skin infections. They wore T-shirts that contained more holes than material and ragged shorts, grey or faded red, with elasticated top. The owner waved them away so that they would not crowd him and block the daylight.

Each boy turned up the sack he had carried since early morning. Thick particles of dust flew up in the sunshine as the waste metal, paper, metal, plastic – pitiful gleanings of their eight hours on the road – fell from the container. The owner took out a pair of tarnished metal scales. Most of the contents of the sacks would have been familiar to Mayhew – ancient springs, a twisted piece of iron, old clothing, shards of glass, half a bottle, some nails, a newspaper. He would not have recognised plastic, and he might have been astonished to find that there is no bone – this is appropriated directly from butchers and slaughterhouses. The owner pays 10 taka a kilo for plastic, 6 taka for metal other than iron, 7 for iron, one and a half taka for broken glass, 2 taka for bottles, 3 taka for newspapers. Fifteen or sixteen children depend upon him for their daily payment.

Mayhew described the life of the mudlarks, known by this name

from being compelled, in order to obtain the articles they seek, to wade sometimes up to their middle through the mud left on the shore by the retiring tide. These poor creatures are certainly about the most deplorable in their appearance of any I have met with in the course of my inquiries. They may be of all ages, from mere childhood to positive decrepitude, crawling among the barges at the various wharfs of the river ... The mudlarks collect whatever they happen to find, such as coals, bits of old iron, rope, bones, and copper nails that drop from ships while lying or repairing alongshore ... the coals that the mudlarks find, they sell to the poor people of the neighbourhood at 1d per pot, holding about 14lbs. The iron and bones and rope and copper nails which they collect they sell at the rag shops ... At one of the stairs in the neighbourhood of the pool, I collected about a dozen of these unfortunate children; there was not one of them over twelve years of age, and many of them were but six. It would be almost impossible to describe the wretched group, so motley was their appearance, so extraordinary their dress, and so stolid and inexpressive their countenances. Some carried baskets, filled with the produce of their morning's work, and others old tin kettles with iron handles. Some, for want of these articles, had old hats filled with the bones and coals they had picked up; and others, more needy still, had actually taken the caps from their own heads and filled them what they had happened to find ... On questioning, one said his father was a coal-backer; he had been dead eight years; the boy was nine years old. His mother was alive; she went out charring and washing when she could get any such work to do. She had 1s a day when she could get employment, but that was not often: he remembered once to have had a pair of shoes, but it was a long time since ... he could neither read nor write, and did not think he could learn if he tried 'ever so much'. He didn't know what religion his father and mother were, nor did he know what religion meant. All the money he got he gave to his mother, and she bought bread with it, and when they had no money, they lived the best way they could.

Another was only seven years old. He stated that his father was a sailor who had been hurt on board ship, and had been unable to go to sea for the last two years. His elder brother was a mudlark like himself. The two had been mudlarking more than a year. They gave all the money they earned to their mother. Another of the boys was the son of a dock-labourer – casually employed. He was between seven and eight years of age, and his sister, who was also a mudlark, formed one of the group.

Mayhew comments:

> There was a painful uniformity in the stories of all the children: they were either the children of the very poor, who, by their own improvidence or some overwhelming calamity, had been reduced to the extremity of distress, or else they were orphans and compelled from utter destitution to seek for the means of appeasing their hunger in the mud of the river. That the majority of this class are ignorant, and without even the rudiments of education, and that many of them are from time to time committed to prison for petty thefts, cannot be wondered at. Nor can it even excite our astonishment that, once within the walls of a prison, and finding how much more comfortable it is than their previous condition, they should return to it repeatedly...

I went with Iqbal, a fifteen-year-old worker, to Sadaraghat, which is the central River Launch Terminal in Old Dhaka. The streets here are narrow and congested with cycle-rickshaws, baby-taxis, Tempos, vans, cycles, carts pulled by men and boys, all carrying people and goods to and from the launches that constantly arrive and depart from the terminal. The traffic gets inextricably caught in endless jams, a tangle of metal and humanity, blocking the road, giving rise to noisy disputes and exchanges, a honking of horns, ringing of bells and yelling of voices. People on foot often move faster than the vehicles.

The terminal itself is a long concrete building, with a number of gates leading down to a rough quayside. We pay 5 taka to go through the rusting metal turnstile. From there, a long concrete corridor with a stone floor, from which, at intervals of about twenty metres, wooden walkways lead down to the quay. From here passengers and goods get in and out of the cargo boats, passenger ferries, country boats, steamers and launches, some of them rusted and creaking, others crudely made. A consignment of watermelons or green coconuts, a load of newly-dyed cloth, headloads of fish, mangoes or papaya, people arriving from Khulna or Barisal, create a chaotic and colourful perpetual movement. On the walkways, and on the quayside itself, crowds of vendors selling biscuits, bread, bananas, soft drinks, for the refreshment of workers and travellers. Some people simply lie on the wooden walkways, sick, exhausted, unemployed, simply resting – it is often difficult to tell which. Most men are dressed in a check *lunghi* or ancient trousers and faded shirts, many are barefoot. A number of handicapped children are begging; some with twisted limbs, others brain-damaged.

Crowds of children live here, on the terminal, perhaps as many as 200. It is significant that so many abandoned, lost, runaway or discarded children inhabit travel termini – the railway, the bus stations, the river terminal. These places are symbolic of the social and emotional limbo in which they exist in a country whose official ideology sees the family as the only secure locus of being. It is as though the children are waiting to go back to where they belong or to move on to a destination of which they are uncertain.

Many of the children here are separated from their families; orphans, strays and others who have failed to be caught up in the only welfare nets available in Bangladesh, which are formed by flesh and blood, the lattice of outstretched arms. No one else takes responsibility for them, these damaged, sometimes ruined, children, whose destiny it is to be soon swallowed up in the undif-ferentiated anonymity of adulthood – a status that will rob them even of the shallow sympathy that their status as lost children may briefly earn them.

We sit on one of the concrete benches inside the terminal, and soon we are besieged by a crowd of children who press in on us, a wall of dusty flesh, a breath of unwashed young bodies, a strangely feral jumble of thin arms and legs and burning eyes. When they speak, a curious composite form of existence emerges – almost as though the individual stories are not the biographies of individual children but part of a collective epic of dispossession and loss. They are like the chorus of some terrible tragedy, orphans of war, the war of the rich against the poor, casualties of economic catas-trophe, unclaimed human leftovers of a maldevelopment which has stranded them on these abandoned sites between coming and going, arrival and departure, denying them both origin and desti-nation.

Therese Blanchet is surely exaggerating when she says of the street children of Bangladesh that they have the greatest space to create a world of their own – the freedom of the streets is intoxi-cating, and once a child has known it, he or she cannot go back to the constraints of family life. 'These children live in a heroic world, where the ability to survive is attributed to their own wits and nobody else's.' Similarly, she says that children living by *dhanda*, illegal activity, exult in the freedom this gives them. 'When they speak of running away from poverty, unloving families, heavy work and onerous study, they are offering a critique of the *samaj* or society.' Yet most children had scarcely chosen their role as social critic: if many tell a heroic tale, this is perhaps to hide the shame and pain of what often turns out to be abandonment, neglect and cruelty. 'I ran away' is a more acceptable version of autobiography than 'It was impossible for me to live in an unloving family.'

Ismail is about twelve. His family is from Barisal, but he does not know where. He says he was born here on the terminal. He has one sister, but never sees her. He lives and works here on Sadaraghat. He is a coolie, earning 20 to 25 taka day; he sleeps on the concrete benches, he has no other clothes than his black trousers, shiny with wear. Ismail eats twice a day, bread or rice from a stall outside, and works from seven in the morning until ten o'clock at night. The police sometimes give the children trouble by beating them and even, sometimes, taking from them the money they earn.

Many of the children carry plastic bowls, which they use for begging. Some are wearing the empty upturned container on their head. Yurub Ali is eleven. He has no family and has been here for three months, begging. He was studying at a *madrasa*, a religious school, but when his mother died, his father abandoned him, so he came here, where he knew other children lived. Sometimes he works as a coolie, but if there is no work, he begs. Even in these impoverished places, the very crowds ensure that there is always someone who will give the necessities for survival – a chappati, a dish of unfinished rice, some rotting fruit, a few taka. Yurub Ali wears a little checked *lunghi* and a blue shirt. He is very thin, and his eyes look into mine, insistent, appealing.

Beauty is a small girl with close-cropped hair and wearing a faded ragged dress. She says that her father and mother are dead, and that she has lived on Sadaraghat for three years. She is ten. She helps the owner of a vegetable stall and earns 20 to 30 taka a day. She sleeps on the dusty stone surface of the terminal floor. She eats at a stall outside. Beauty has no relatives, but she says the children look after one another. With a simple gesture, she says, 'These are my family.' Although each child is competitive, looking for any opportunity to survive, there is at the same time a network of protection between them; and the readiness with which Beauty uses the word 'family' suggests that even the most outcast reconstitute themselves in the image of the dominant institution. The abandoned children recognise in each other the only resource they have. Beauty came to Dhaka with her parents, but they both died within a short time. She has no brothers or sisters. She would like to work in a house. She has heard of children who work as housemaids and have a mat to sleep on under a roof. Such a modest ambition, a longing for such exploitative work, but to her, a young girl stranded in the constant menacing presence of travelling strangers, most of them men, it must seem like deliverance.

As we leave the Sadaraghat, we are caught in a web of empty entreating arms that will not let us go.

The differences between Mayhew's waifs and the children on Sadaraghat are also significant. Their expressions are far from the 'stolid inexpressive countenances' of the London ones; it is their smile, that collective and poignant patrimony of the South Asian poor, their darting mobility and energy that strike the observer. Mayhew's children were for the most part still living in families, even though one or both parents may have been dead. They were not the most derelict of London's children, whereas these are the most abandoned of Dhaka's. The London children had little sense of religion, yet none of the Bangladeshis was unaware of their own faith. Indeed, it is clear that, whatever its disadvantages, Islam serves as a more effective inhibitor of crime and violence in contemporary Dhaka than Christianity did in early nineteenth-century London.

Yet whatever the cultural or social differences, the exclusion, the indifference of the authorities towards their fate remain the same. Mayhew's observation that prison might be more comfortable for them is an expression of his own social prejudice. The Dhaka children are far less criminal than their London equivalents might have been; and, in any case, they will do anything to avoid the police, who are their sole point of contact with the legal system, such as it is – the police regard the poor as an unofficial means of augmenting their income. They beat, mistreat, wrongfully arrest and steal from them with impunity; and to fall into their hands is a fate even worse than living on the mildewy concrete of the river terminal. The ingenuity with which some of Mayhew's urchins had made work for themselves is repeated in the children who find a niche in the urban economy of Dhaka.

Dhaka is a very flat city, but in the Sutrapur area there is a slight incline in the road; not very steep, but the closest thing to a hill. At this point, the cycle-rickshaw drivers usually dismount to push the vehicle up the slope. A number of boys work here, helping to push the rickshaw with its passengers up the gradient, which saves the driver the effort of stepping down. As the rickshaw approaches the hill, some barefoot boys take it in turns to shove; some of the smaller ones must give it all their strength.

Rayhan is nine and he works in partnership with his eleven-year-old brother. Together, they make up to 60 taka a day, mainly from passengers who sometimes give them 1 or 2 taka for their help. Some of the drivers also give them a few taka. Rayhan is a small thin child; it is impossible to measure the effort that goes into his labour. Their father is also a rickshaw driver, and this is what Rayhan expects to do when he is older. Both boys go to a government school from ten to one o'clock and can write their name. Rayhan has been working here for a year.

Shahabuddin does the same work. He is thirteen and expects to earn 25 to 30 taka for working all day. He used to go to school, but his father is now blind and cannot work. He has one brother who is fourteen and drives a cycle-rickshaw. His parents migrated from India 'long ago'. Shahabuddin wants to become a mechanic or engineer. His family lives about eight kilometres from here, and he travels by bus. Since this costs 10 taka a day, his earnings are reduced to 15 taka. He eats once during the day and again when he gets home. He gives all the money he earns to his mother.

Anwar is ten. We were sitting in a small restaurant drinking tea on Topkhana Road. It was mid-afternoon, and Anwar came in for a metal jug of steaming tea. He works in a government office, running errands, fetching tea, *paan*, water and cigarettes for the workers in the Income Tax Office. He works from nine to five or six in the evening and is paid 400 taka a month (about US$ 8). Anwar wears a grey shirt, grey shorts, no shoes. His brother and two sisters are all younger. He goes to school for one hour from seven to eight in the morning before starting his day's work. The school is run by a Japanese NGO. He lives in Mirpur, about ten kilometres away. Anwar's father drives a baby-taxi and his mother is a cleaner in the Tax Office. They come from Kustia, 200 kilometres from Dhaka. It is Anwar's ambition to get a job in the Tax Office when he grows up. He is a small child, nervous that he will be scolded if the tea is cold by the time he returns to the office. It seems to strike no one as odd that a government ostensibly committed to the abolition of child labour nevertheless employs children of ten in its own service.

Chapter Thirteen

If we approach the present-day experience of child workers in South Asia and compare it with the images from the nineteenth century, the parallels are no less remarkable.

One of the most common employments for young girls in Dhaka is in domestic service. You have to be watchful even to notice them, the little maids of Dhaka. They can sometimes be seen, shadows on verandahs, pinched faces patterned by the ornamental grilles that are also prison bars.

There are, it is estimated by SHOISHOB, an organisation working with child domestic labour in Bangladesh, between 250,000 and 300,000 of these young girls in Dhaka alone. You often hear middle-class women say of their servant, 'Oh, she is just like my daughter.' Rarely are the children heard to say of their employer, 'Oh, she is just like my mother.'

In 1999 SHOISHOB published a survey which covered more than 10,000 middle-class households. In these, nearly 8,000 resident servants were counted, of whom 2,500 are minors. More than 80 per cent of them are girls. The age of these maids show two peaks – at ten and eleven, and again at fourteen or fifteen. There are good reasons for this. Some families prefer pre-pubertal maids, since they are less likely to attract the sexual attentions of the men of the household.

These children are 'bandhu', that is 'tied' servants, who live with the employer's family. This is one of the most secret and inaccessible forms of child work, enfolded within the privacy of individual households. It is also, for the children themselves, one of the most hazardous occupations, not excepting the flower sellers who weave in and out of the moving traffic, the workers in dangerous garages and repair shops, the workers in plastics, chemical and shoe factories: it is simply that for the little maids the hazards are different.

The servants are children of some of the poorest families. Their earnings are often negligible. A majority of those surveyed by SHOISHOB receive between 100 and 200 taka a month ($US 2–4). The second highest number receive no pay at all. They are fed and housed, which has two advantages for their impoverished families. It represents for them one less mouth to feed, and it is also believed – not always correctly – that shelter in a middle-class household will protect the future marriageability of the girl. That is to say, the environment of the well-off family is thought by the slum-dwellers to be a safe place, safer than the slum and village neighbourhoods where they live.

This perception is often mistaken. 'We know that large numbers of child servants are abused', said Helen Rahman of SHOISHOB, 'not only beaten, but also abused sexually'.

As far as older girls are concerned, the woman of the house sometimes colludes with a sexual relationship between her husband and the servant. They may feel it preferable that their husbands' extra-marital sexual activities should be kept under one roof. They are less likely to look for consolation outside the home. If the son of the household should form a relationship with a servant, it is perhaps unlikely to be viewed with the same complaisance.

This is confirmed by the child domestic workers whose lives were investigated by Therese Blanchet. She says that girls between the ages of eight and twelve are preferred, 'before their eyes are open', as the employers say; that is, while they obey and do not steal. Domestic work is seen also as a good preparation for the girl's destiny as wife. The relationship reinforces the ancestral master–servant, patron–client relationship. Blanchet found, in the eighty servants whose situation she reviewed, that as many as 76 per cent received no wages, although almost as many were given clothing. Employers are seen as guardians, which implies moral guardianship also.

The work is very exploitative and the pattern of work similar for the great majority. Even those who regard themselves as liberal towards their maidservants often do not perceive the demands they make on them. 'It is a habit of mind', Helen Rahman said. 'They believe they cannot live without servants. It is this attitude which must be challenged.'

I was struck by the truth of this a couple of days later, when I was talking with a highly educated professional woman whose two daughters were in the USA, both married to Americans. She said she could not even think of settling in the USA because she has been spoiled by having everything done for her by servants; a privilege in Bangladesh, which she knows she could not find in the United States.

Domestic labour is commonly called *shajao dawa*, help, or *choto kaj*, small work; a designation calculated to reconcile the employers to the burdens they impose on the children. 'Oh, she does nothing very much', is commonly heard. In practice, their presence is indispensable to the running of the household.

The child domestics are on call up to sixteen or seventeen hours a day. They must be up in the morning before anyone else in the household. They sweep the floor, wash their face, prepare breakfast for whenever the householders and their children require it. They clean and cut the vegetables, prepare and sometimes cook the meal. They look after the house, attend to the needs of small children, wait on visitors, and if there is a party they are expected to serve all comers. They must wait to eat until the family has finished; and although they most probably get better food than they would at home, it is inferior to that which the employers eat. The servants are always the last to go to bed, after they have cleared away the evening meal, washed the vessels and set everything ready for the morning.

In Dhaka, there are three options for poor young women. All involve clothing. They can make clothes, like the 800,000 who work in the garment factories. They can wash clothes, like the hundreds of thousands of maidservants, or they can take their clothes off, whether as sex workers to service strangers or as youthful brides. It is far from evident which of these is the least onerous fate.

Helen Rahman believes it is not helpful to attack child domestic labour from a high moral position. 'This only hardens attitudes', she says. SHOISHOB runs schools for about 3,000 child servants in Dhaka, although, of course, these reach only the children employed by the most enlightened sections of the middle class who are prepared to release them for two or three hours a day. The schools are open for two hours in the afternoon, from three to five o'clock, when domestic duties are generally at their lightest.

We visited one centre at Mirpur in north Dhaka. A ground-floor verandah in a substantial house, given for the purpose by a prosperous family. Bamboo mats had been spread on the concrete, and there were twenty children, aged from about ten to fifteen, most of them girls, with three boys present. They concentrate on their slates with the greatest application, studious, obedient. This may be read either as evidence of the discipline of the labour to which they have long been accustomed or of the enjoyment they take from the only activity in the day in which they are the focus of attention. Most of the children state that the afternoon – and that means school – is the best moment of the day.

The ambition of most of them is to become 'job-holders', that is government employees. Some say that they would like to be a doctor or a teacher. One of the least understood effects of life in middle-class families for the children of the poor is how their identity and sense of self may be impaired by continuous exposure to contact with the alien values of the households in which they serve; particularly if, as is often the case, they change jobs frequently. They do not do this by choice. Their parents or guardians decide when they will be moved, whether to improve their income or to place them in a 'better' household. The parents may come to resent the growing involvement of their child with the employing family and may move her without any reference to her wishes. Therese Blanchet says:

> A number of servants interviewed by Anisa Zaman had their names changed by employers, an intervention much resented by the children, but one which could not be objected to. 'I have a beautiful name, but it was changed to another one,' said a ten-year-old girl. In one household, a succession of boy-servants were all called 'Abdul'.

The same experience was familiar in Britain even within living memory. The mother of a school friend of mine was called Adelaide. When she went into service at thirteen, she was told, 'I shan't call you that. Adelaide is far too pretty a name for a servant. I shall call you Ada.' In Britain, it was usually on grounds of class that the names of servants were altered, if their parents had had the temerity to name them after a member of the Royal family, for instance. Therese Blanchet observes that child brides, child prostitutes and child servants all go through a de-naming process, which is depersonalising; calculated to obliterate past history and selfhood and to impose a new identity.

I asked the children in the Mirpur school what they dreamed about at night. One child said she dreamed of the holy man who died in the family she served, and some mentioned the TV series of Sinbad, which many are allowed to watch; but most spoke of dreams of the families from which they have been arbitrarily – it seems to them – separated. Many must feel this as a punishment for wrongdoing they cannot understand, for mysterious sins they might have unwittingly committed; sentiments which exile in their unchosen place of residence often serves to confirm.

Confusion over identity was familiar to young servants in Victorian England. Many assumed the language and manners of their employers; some became snobbish and déclassé, and thought themselves too good for the work of bringing up their own family. The mother of one man I know worked as a cook: she bitterly resented the rest of her life, the 'uncouthness' of her husband, and dreamed of returning to the life of which she accused him, unjustly, of robbing her. Indeed, penetration of servants by the values of their social superiors was probably a significant means of social control throughout the nineteenth century. Many basked in a borrowed lustre from those they had served, named their children after theirs, treasured their years of living in houses from which their own life represented a permanent coming down in the world.

Overwork, extreme exhaustion: many servants are expected, even in their 'free' moments, to look after young children. They are often left to mind the apartment when the mistress is out. Few girls are allowed to go on errands outside the home. They remain captive in the building. One girl said, 'I stand on the verandah when there is nothing to do, because there is nowhere else for me to go.'

Dhaka is full of these silent, unobtrusive waifs, without whose ministrations the lives of privilege would be laborious indeed. Yet their employers spend much time complaining of their slowness, their dullness, their unreliability and inability to follow simple instructions. It is astonishing how unnoticed evils may remain, how taken for granted they may be, while they serve the interests of the powerful.

Helen Rahman believes that there are two possible ways to improve things for these exploited children. One is to shame those people who depend unreflectingly upon the services of their choiceless child servants; and the other, to claim recognition for domestic work as the labour it is. 'If the girls had contracts, clearly stated duties, due definition of what is expected of them, this would put an end to some of the worst abuses. As it is, most girls have no idea of the limits to their duties. They accept all orders, take as given the demands made upon them.'

At the same time, a degree of perfection which is not expected in any other form of employment is demanded of them. If a child breaks a dish or an ornament, she will be scolded and, possibly, punished. She will be told only of the things she does wrong; almost never praised for the patience and diligence with which she performs her daily duties.

As they grow towards adolescence, many employers claim that their servants become 'spoiled'. Therese Blanchet says that they assume that servant girls come to the age of 'buddhi' or understanding earlier than their own children. At the age of fifteen or sixteen, maidservants who have not been paid start to ask for a salary. This is when employers see them as 'spoiled'. This is a key word, says Blanchet, in understanding Bengali childhood.

The West perceives growing up as to be seen in acts of rebellion or disobedience. Bangladeshi society does not recognise the morality of adolescent rebellion. This is important for the administration of justice. Children who are perceived as having lost their innocence are assimilated to adults. Adolescents who commit crimes are said to 'understand', and therefore to have the same responsibility as adults.

In many households in Bangladesh and India, the dishonesty of these older servants is a major preoccupation. 'These people are crooks', I remember one householder solemnly admonishing me. 'You cannot trust them.' She did not see the pitiful salary her servants received as an element in their 'criminality'. The employers lock the fridge, weigh the sugar, measure the coffee in the jar, convinced that the servants will be the ruin of them with their extravagance, untrustworthiness or criminal intentions.

In 1851, Mayhew exhibited the same certainty about the unreliability of servants. He asserted:

> Many felonies are committed by female servants who have been only a month or six weeks in service. Some of them steal tea, sugar and other provisions, which are frequently given to acquaintances or relatives out of doors. Others occasionally abstract linen and articles of wearing apparel, or plunder the wardrobe of gold bracelets, rings, pearl necklaces, watch, chain or other jewellery, or of muslin and silk dresses and mantles, which they either keep in their trunk or otherwise dispose of...

It was significant that of the twenty children in the Mirpur school more than three-quarters of them have come directly from the country. This rouses echoes of Victorian Britain, when unspoilt country girls were preferred as maidservants in the cities because they had not been ruined by urban life. Precisely the same justification and the same rationalisations are at hand in another culture, another climate, 150 years later: the same frightened children, apprehensive, docile, waiting to be moulded by whatever values and expectations the receiving families choose to impose upon them.

The small maids of Dhaka are perhaps as close as the contemporary world comes to slavery. It comes, therefore, as no surprise that accounts from the eighteenth and nineteenth centuries of domestic – and often sexual – slavery of children and young people on the plantations in the Caribbean prefigure in so many particulars the experience of the child servants of Bangladesh.

Walvin states:

> By 1700, black domestics had become the norm. One man remarked in 1708 that 'the handsomest, cleanest [black] maidens are bred to menial services'. Planters and slave-owners could of course delude themselves, and could equally be deluded by the slaves, that the bonds between slaves and owners were more affectionate than they were in reality. Domestics ... shared their owners' lives more intimately than did other slaves. They had access to the most private of plantocratic possessions: foodstuffs (which they sometimes poisoned), medicines (which

they sometimes used to harm or kill the whites) and above all, their children ... Most slave-owners employed someone as a domestic. In Barbados in 1788 it is thought that upwards of 25 per cent of the overall slave population were employed as domestics. In Bridgetown, almost 70 per cent of all females in the town were domestic servants ... Domestic slaves had their own stratagems and wiles for coping with their lot. It was universally recognised that they were cunning; they stole and they lied (when caught they were soundly whipped and punished in other ways). They, like slaves in general, were thought to be impervious to any discipline short of the lash or physical punishment. 'To kindness and forbearance, they return insolence and contempt' ... The most troublesome aspect of life was the flagrant and often aggressive sexual approaches from local white men. The few who wrote diaries openly spoke of their sexual adventures from their early years. William Byrd II of Virginia recorded in 1720, 'I felt the breasts of a Negro girl which she resisted little.' Similar scenes were enacted among the sons of Robert Carter, another famous Virginia planter. His oldest son Ben was accused by a younger brother of taking a young slave girl 'into your stable and there for a considerable time lock'd together'. Later, Ben himself was suspected of breaking into a house in order to 'commit fornication with Sukey' (a plump sleek Negro girl about sixteen).

Walvin reports that 'domestics were expected to rise before dawn, and still be on hand long after dusk to serve, feed their owners, then prepare the house for the following day ... Plantation mistresses who reared domestic slaves from an early age often felt very attached to them; they spoke or wrote about having reared them as their own children.'

When Chateaubriand, fleeing the Revolution in France, sailed up the Chesapeake Bay, the first person to meet him was 'a Negress of thirteen or fourteen, practically naked and singularly beautiful ... I gave my silk handkerchief to the little African girl: it was a slave who welcomed me to the soil of liberty' (Walvin 1993).

Chapter Fourteen

The early nineteenth century in Britain and late twentieth in South Asia present aspects which are, at the same time, astonishingly similar and incomprehensibly different. Yet despite the technological changes at which the world marvels, the poor remain poor through time in the same way. (The rich do too: conspicuous consumption, display and ostentation are not recent phenomena. Perhaps the greatest novelty among the rich in recent times is their desire not to draw too much attention to themselves, for fear of crime. On occasions when they go out into the city, it is not unusual now for people to 'dress down'. Women remove their jewellery, their watches, rings and bangles, before venturing to the shops; they wear simple clothes to travel in a cycle-rickshaw, train or bus. Only in the fortified enclaves of home and at private parties do they deploy the full regalia of their wealth and privilege.)

The impression we gain of the great cities of South Asia is a curious sense of time foreshortened: the filth and impoverishment of the early nineteenth century mingle with some of the amenities of the modern world, including drugs that have spectacularly lengthened life expectancy, as well as material innovations – especially TV, refrigerators and electric fans – in some of the most wretched slums. The consolations of the poor – tobacco, sugar, liquor (in Bangladesh, home-made 'Bangla moth', often adulterated and highly dangerous), forbidden drugs, laudanum and other opiates – remain largely the same; only many of these are now brought to them by transnational companies or by the equivalent illegal organisations which distribute crack, brown sugar, *shabu* or cough mixture, Phensidyl, illegally imported from India, glue or 'solution' to the young and most impoverished. While the dust and pollutants from vehicles are an addition to the unhealthiness of the atmosphere, canals, ponds and waterways charged with chemicals and decaying matter are little different from those described by Engels. The exploitation of labour, conditions in basic units of manufacture – especially in spinning, weaving and making cloth – have varied little; and while malnutrition, infant mortality and low birth weight remain, they have been mitigated by medicines and drugs that simply did not exist in Victorian England.

Mayhew's evocation of the Saturday night markets of London might have been taken from contemporary Dhaka.

The scene, in these parts has more of the character of a fair than a market. There are hundreds of stalls, and every stall has its own one or two lights; either it is illuminated by the intense white light of the new self-generating gas-lamp, or else it is brightened by the red smoky flame of the old-fashioned grease-lamp. One man shows off a yellow haddock with a candle stuck in a bundle of firewood; his neighbour makes a candlestick of a huge turnip, and the tallow gutters over its sides; whilst the boy shouting, 'Eight a penny, stunning pears!' has rolled his dip in a thick coat of brown paper that flares away with the candle. Some stalls are crimson with the fire shining through the holes beneath the baked chestnut stove; others have handsome octahedral lamps, while a few have a candle shining through a sieve; these, with the sparkling ground-glass globes of the tea-dealers' shops, and the butchers' gaslights streaming and fluttering in the wind, like flags of flame, pour forth such a flood of light that at a distance the atmosphere immediately above the spot is as lurid as if the street were on fire.

Then the sights, as you elbow your way through the crowd, are equally multifarious. Here is a stall glittering with new tin saucepans; there another, bright with its blue and yellow crockery, and sparkling with white glass. Now you come to a row of old shoes arranged along the pavement; now to a stand of gaudy tea-trays; then to a shop with red handkerchiefs and blue checked shirts, fluttering backwards and forwards, and a counter built up outside on a kerb, behind which are boys beseeching custom. At the door of a tea-shop, with its hundred white globes of light, stands a man delivering bills, thanking the public for past favours, and 'defying competition'. Here, alongside the road, are some half-dozen headless tailors' dummies, dressed in Chesterfields and fustian jackets, each labelled 'Look at the prices' or 'Observe the quality'. After this is a butcher's shop, crimson and white with meat piled up to the first floor, in front of which the butcher himself, in his blue coat, walks up and down, sharpening his knife on the steel that hangs to his waist. A little further on stand the clean family, begging; the father with his head down as if in shame, and a box of lucifers held forth in his hand – the boys in newly washed pinafores, and the tidily got-up mother with a child at her breast. This stall is green and white with bunches of turnips – that red with apples, the next yellow with onions, and another purple with picking cabbages. One minute you pass a man with an umbrella turned inside up and full of prints; the next you hear one with a peepshow of Mazeppa, and Paul Jones the pirate, describing the pictures to the boys looking at the little round windows. Then is heard the sharp snap of the percussion-cap from the crowd of lads firing at the target for nuts; and the moment afterwards, you see either a black man half-clad in white, and shivering with cold, with tracts in his hand, or else you hear the sounds of music from 'Frazier's Circus' on the other side of the road ... Such, indeed, is the riot, the struggle and the scramble for a living, that the confusion and uproar of the new-Cut on Saturday night have a bewildering and saddening effect upon the thoughtful mind.

The same scenes spring to life in the market of Bangabazar or around mosques and at the park entrances. At dusk, the kerosene lamps, oil lamps, candles and bare electric bulbs create a mixture of coloured lights which flicker over a different produce, but with the same kaleidoscopic sense of colour and movement: a candle in the green shell of a soft coconut, the yellow flare of kerosene on pyramids of mangoes, papaya in newspaper, fingers of small bananas, long white radishes, purple onions. The snacks – sticky shapes of sugary sweetness, peanuts, custard apples, cakes and biscuits – attract swarms of flies. The mobile carts on which vendors offer ballpoint pens, hair-slides, brooches, are similar in design to those observed by Mayhew; cheap tracts, second-hand books, belts and buckles, ribbons, bows, bootlaces, candies, toys – windmills, balloons, dolls – all of these would have been familiar; music cassettes, cheap radios, electronic games, plastic cars and aeroplanes may have displaced earlier small luxuries, but the overall atmosphere remains the same.

Even the entertainments – the air rifles shooting at balloons on a board, the hoop-la and the ubiquitous improvised cricket games – have scarcely evolved, although the peepshow has been overtaken by TV and the penny dreadfuls by electronic games. The inventiveness of children in finding a niche in the urban economy, their sharp-wittedness and adaptability are unchanged. They tell the same stories that Mayhew's informants offered with such candour. They left home because of an unkind stepfather, because they were hungry, mistreated, because they were rebellious or curious, because they were abandoned by their father, because siblings or relatives had been weakened by diarrhoea or cholera. Where the starvelings in Mayhew suffered from cold and wet, in Bangladesh violent natural upheavals also contribute to the displacement of people – the river that ate the parcel of land and washed away the village, the tidal wave that drowned the buffaloes, the floods that alter the landscape each year. Yet there is little difference between the waifs and street children, separated by six generations or more, and by culture and religion.

Those who speak sagely of the differences between ways in which countries in a unipolar world develop in their own diverse cultural ways exaggerate. Social injustice has the same stark materiality everywhere. People are deprived of the same things, in the same way. Necessities are unavailable, not because of natural scarcity, but because they lack the power to buy in the abundant markets. The goods they require for survival are withheld by the rich, whose lives this monopoly does not improve or ornament, but the absence of which certainly robs the poor of nutrition, health and education. Whatever cultural embellishments distinguish the orphans of the Dhaka railway and river terminals from the mudlarks of nineteenth-century London, the hunger, the creativity, the determination to survive, the ragged clothing and resented dereliction make for far greater similarities.

In 1889, Charles Booth described the London slums a generation after Mayhew. He evoked one street that was

just wide enough for a vehicle to pass either way, with room between curb-stone and houses for one foot-passenger to walk; but vehicles would pass seldom, and foot-passengers would prefer the roadway to the risk of tearing their clothes against projecting nails. The houses, about forty in number, contained cellars, parlours, and first, second and third floors, mostly two rooms on a floor, and few of the 200 families who lived here occupied more than one room. In little rooms no more than eight feet square would be found living father, mother and several children. Some of the rooms, from the peculiar build of the houses (shallow houses with double frontage) would be fairly large and have a recess six feet wide for the bed, which in rare instances would be curtained off. If there was no curtain, anyone lying on the bed would perhaps be covered up and hidden, head and all, when a visitor was admitted, or perhaps no shyness would be felt. Most of the people are Irish Roman Catholics getting a living as market porters, or by selling flowers, fruit, fowls or vegetables in the streets, but as to not a few it is a mystery how they live. Drunkenness and dirt and bad language prevailed, and violence was common, reaching at times even to murder. Fifteen rooms out of twenty were filthy to the last degree, and the furniture in none of these would be worth 20s, in some cases not even 5s. Not a room would be free from vermin, and in many, life at night was unbearable. Several occupants have said that in hot weather they don't go to bed, but sit in their clothes in the least infested part of the room. What good is it, they said, to go to bed, when you can't get a wink of sleep for bugs or fleas ... The passage from the street to the back door would scarcely ever be swept, to say nothing of being scrubbed. Most of the doors stood open all night as well as all day, and the passage and stairs gave shelter to men who were altogether homeless. Here the mother could stand with her baby, or sit with it on the stairs, or companions would huddle together in cold weather. The little yard at the back was only sufficient for dustbin and closet and water-tap, serving for six or seven families. The water would be drawn from cisterns which were receptacles for refuse, and perhaps occasionally a dead cat ... gambling was the amusement of the street. Sentries would be posted, and if the police made a rush, the offenders would slip into the open houses and hide until the danger was past. Sunday afternoon and evening was the heyday of the street. Every doorstep would be crowded by those who sat or stood with pipe and jug of beer, while lads lounged about, and the gutters would find amusement for not a few children with bare feet, their faces and hands besmeared, while the mud oozed between their toes...

There are slums in Dhaka which are also dilapidated buildings: concrete structures, with bare flights of stairs, landings and dark corridors, with single rooms which are home to a whole family. In Old Dhaka, there are also rough constructions to which a second or third storey have been added and which are reached by narrow walkways of stone. Yet others are rooms built around stony courtyards, rough brick with tin roofs, where goats, hens, dogs and cats come and go at will.

There may be fifteen or twenty such dwellings, with children playing recklessly around carts and cycle-rickshaws. The chief item of furniture in each one is a bed, which often fills three-quarters of the living space; a cupboard or a trunk contain such modest valuables as a few teacups, a thermos flask, a watch and a little jewellery. Changes of clothing are on a string across the room or above the bed. There will be a small *chulla* for cooking, which in dry weather will be placed on the path outside the room, but in the monsoon, when the roads are often knee-deep in mud, the smoke will add to the pollution and humidity of the interior. There is often a single water tap serving a score of rooms, and the water supply is intermittent. There are many illegal electricity connections, usually subcontracted by someone who has a legal connection. Inside the warrens of rooms, concrete reddened by the spitting of betel nut, the sounds of a humanity living pell-mell: a man and woman quarrel, young girls sing, boys laugh, babies cry, and everywhere there are crowds of children, the elder carrying the younger, some of the older girls with slightly deformed hips where they have carried their siblings ever since they could lift them.

Most of the slums in South Asia, however, are not like this. They consist of bamboo frames, with *chetai* (woven bamboo) walls, and sometimes roofs, although many people have replaced these with corrugated metal. Some have been built by the occupants themselves, who bring skills in self-provisioning construction from the villages.

The slum families in South Asia have a longer memory of self-reliance than their counterparts in nineteenth-century London and Manchester. In Bangladesh, memories of rural self-sufficiency were less comprehensively destroyed than those of the peasantry of Britain. The poorest are perhaps less brutalised than they were in early industrial Britain. In Dhaka a combination of Bengali culture and Islam certainly makes the scenes of slovenliness, squalor, drunkenness and foul language monitored by Mayhew and Booth quite foreign to Bangladesh. Of course, both Mayhew and Booth were deeply affected by the prejudices of their class, but their account of ignorance, desperation and violence no doubt contained much truth. In contrast, while in Calcutta, Mumbai or Dhaka there are examples of terrible indifference by the authorities and epic economic violence, inter-personal relationships among the poor are marked neither by quarrelsomeness nor by a breakdown in civility. Quite the contrary. The low levels of literacy in South Asia should not be taken as evidence of either stupidity or backwardness, any more than a largely literate poor of London by the early twentieth century should be seen as a particular sign of progress. There remains a pre-industrial innocence among the poor of Dhaka, and their very illiteracy may have been an agent in the preservation of an ingenuousness that has not become cynicism, an openness that has not been transformed into sullen suspicion.

What is common to the industrial population of nineteenth-century Britain and to contemporary Dhaka is the closeness of many of the poor to the countryside from which they so recently came. Indeed, the great majority of the young women workers in the garment factories were living in the countryside in the mid-1980s. The population of Dhaka has probably doubled within the past twenty-five years; just as in the first thirty years of the nineteenth century, the population of greater London rose from 865,000 to 1,500,000, while in the next twenty years another million was added. Peter Quennell, editor of *Mayhew's London Labour and the London Poor* (1984) says:

As the population thickened, so did its occupations grow more and more miscellaneous, its character more amorphous. Parasites fastened on parasites; the refuse and leaving of one class helped, literally as well as figuratively, to provide a means of livelihood for the class immediately beneath it; and while the poor but 'respectable' members of commercial society, the clerks and small employees, tended to gravitate towards the pretentious gimcrack suburbs, rag-pickers and pedlars drifted into its noisome central slums, into one or the other of the 'rookeries', clusters of dilapidated houses...

The slums of Dhaka, many of them built around ponds and stretches of water, look more like rural villages than city slums; only the accumulation of uncollected rubbish, the foraging animals, the detritus of people living close together and the proximity of industrial areas to the living places create an atmosphere reminiscent of Engels's evocation of Lancashire in the 1840s. Add to the pollution of the nineteenth century the remains of indestructible plastic goods and synthetics, and the mixture is somewhat altered; but it is largely in such familiar places that scores of thousands of children live and work.

Chapter Fifteen

One of the most obvious but least commented areas of child work is how brief the time of child worker is, even though, to an individual child, each day may seem interminable. Children are swiftly absorbed into the anonymity of adulthood, and any tenderness and consideration extended to children are withdrawn as soon as they become adults. This itself is a shifting frontier. Adulthood is often thrust upon children who become responsible for siblings after the death of a parent, which might mean any time from eleven or fifteen; for some, domestic workers, factory workers, it is even earlier.

Very little is known about the influence that child work has upon the adults they become. Many children who have worked when young become highly skilled and command good wages, but this depends upon the kind of training they have received during childhood. Those who have acquired an enduring ability rarely want in later life – many children who worked at lathes, as mechanics, in garages, repairing cycle-rickshaws or baby-taxis, even as operatives in factories, can often transfer what they know to some similar employment. Even in the garment factories, children usually rose from being helpers to operatives, and some later became supervisors. In this sense, work offers forms of instruction that school cannot – the rote-learning, monotony and lack of stimulation in many schoolrooms does little to prepare children for later livelihood; and this is testified to by the large numbers of unemployed graduates in India, Bangladesh and many other countries in the South. Many have undergone a change of consciousness which has made them aware principally of the raised status which their diplomas and certificates bestow – it does not actually produce work.

Too rarely do we gain an insight into the later life of those who worked as children. Karim comes from Satkhira on the edge of the Sunderbans, and he now lives in a slum colony in Khulna, the third city of Bangladesh. He is a small wiry man, the oldest of ten brothers. His family had five acres of land. As a child, from the age of about nine he worked preparing the ground, sowing, planting, weeding and cutting the harvest. He looked after cattle, gathered and cut fodder, looked after buffaloes and goats. He left home together with three other brothers when farming became too costly to employ them. The amount of fertiliser required to maintain the yield increased fourfold in ten years, and the children had to earn the money for the agricultural inputs.

Karim is now a domestic worker with fourteen families in a residential (i.e. middle-class) area of Khulna, earning about 1,600 taka a month. Of this, he sends 600 to the village, where he has a boy aged eight and a girl aged ten. He has built a new house with the money he has remitted. Money from remittances is all that keeps many villages all over the subcontinent from falling into total dereliction.

He now says all the strength of his life has been used up. 'I have worked so hard, my body feels drained. After living in slums for fifteen years, working so many hours each day, there is nothing left in my body. The flies and mosquitoes keep us awake in summer, the cold in winter. There are only two taps for 500 people. In the rainy season it is a sea of mud.' Karim has at times worked for as long as four days with only two hours' sleep. He looks old, thin; his body wasted by overwork and self-denial. He is twenty-nine.

Anwar worked from the age of eight, helping in the fields of Faridpur in Bangladesh. At thirteen, he went to Narayanganj, on the outskirts of Dhaka, to work with his uncle. The uncle had a small shop selling vegetables. Anwar was given food and a place to sleep and 50 taka a month (one dollar). At fourteen, he joined a garment factory as a helper, making T-shirts. He came into Dhaka, joined another factory, but became homesick and went back to his village. He was workless and returned to Dhaka, this time to work as a cycle-rickshaw driver. He paid the owner 45 taka for a twenty-four-hour hire of the vehicle. He could earn up to 100 taka a day, but it was exhausting work. Sometimes he was cheated by the passengers. They would ask him to wait while they fetched money to pay him and then disappeared. Later, he abandoned this job to sell chickens with his cousin.

Anwar eventually met someone who got him a job as peon in the office of an NGO. In fact, much work is like this – if you meet the right person, if you have a friend who speaks for you or an acquaintance tells you there is a job going. A vast informal network of contacts determines who will and will not get regular work. It was very much like this too in the industrial period in Britain.

For Anwar, his present job is a great achievement, which far exceeds his dreams. He keeps the office clean, makes tea, sets out chairs for meetings, serves visitors, runs errands; simple routine work, which he takes very seriously. He is now earning 2,000 taka a month. The work is not arduous. He can scarcely believe his luck. Anwar is capable and intelligent, even though he does not read or write. At twenty-one, he has already participated in almost every possible employment open to a young man, yet he is full of energy and hope. Work as a child taught him to be resourceful, opportunistic, alive to any chance that will provide him with a reasonable livelihood, any possibility of extending his income.

Chipon is now fifteen. He comes from a village about fifty kilometres from Dhaka. His father and a sister died within a few months of each other. His mother works irregularly, according to the season, in the fields of a landowner. Chipon was brought to Dhaka when he was ten by his maternal uncle who found him work in the house of a garment manufacturer; then, clearly having decided he had done his duty, the uncle abandoned the child to the household of the industrialist.

Chipon is sexually abused by this man, and not only by him. When his wife goes away, he brings his friends so that they can also enjoy him. Chipon is a slight, delicate boy; although his features are irregular, his face marked by acne; he has beautiful large eyes, the eyes of a perpetual victim. After two years of domestic labour and sexual abuse, he was also given work in the factory. He now works eight hours a day for 1,200 taka a month (US$ 24). He still fulfils his domestic duties. Other older workers in the factory also abuse him – they take him on to the roof terrace above the factory after dark. Chipon has also learned to work Ramna Park after nightfall, where he makes a few hundred extra taka a week, providing sexual services to strangers. His employer, he says, is unpredictable. Sometimes he is kind, offers gifts of clothing and presents, but at other times he expects him to wait on the family, to carry out all orders, to work in the factory, and to answer the sexual demands of his employer and his friends.

To watch Chipon walking in the flickering lamplight of the park under the dark trees is a sad spectacle. He is small for his age and looks younger than his fifteen years. He looks enquiringly into the faces of passers-by. No one stops him except to make a sexual proposal and to bargain over it, to get the lowest price possible for his services; although some take pity on him and give him more. No one has ever asked him if he is looked after, whether he has a family, whether he has ever been educated, protected, loved. He says that he is looking for love – a need invisible to the urgency of the desires of the adult strangers he services.

If it is difficult for a victimised child like Chipon to articulate his needs or to determine his own conditions of labour, it is almost impossible for girls to protest against their destiny.

Shehana Begum is a trade union leader in the garment industry. In her forties now, she was a child bride who resisted her fate and fought against the life which custom and circumstances imposed upon her.

'My father worked in the time of the Raj on a ship that plied between Calcutta and Dhaka. We were lower middle class, and we owned between ten and fifteen acres of land. After Partition, my father lost his job. He was forced to sell the land little by little, in order to maintain the family. We were seven, four brothers, three sisters.

'During my childhood, we became progressively poorer. An older brother lived in Calcutta, working on a ship, but my father still had to sell more and more of the land to pay for our education. I studied to class ten, and then I had to be married. I was fourteen. My sister, who was even younger, also had to leave school to get married. If we had lived in a town, maybe it would not have happened. But in the village, it is considered time for the girls to be married. In any case, my mother had died, and as girls grow up they are seen as a burden to the family. My grandmother took it upon herself to marry me off. I was taken out of school, quickly married. I had no choice.

'My marriage was not good. Very soon I gave birth to a baby boy. But my husband was idle, he did not work, he had no feeling of responsibility towards his wife and child. He had been married before. I did not know this when I married him.

'He had been a contractor in Dubai when he married his first wife. Then he left her. When my marriage was arranged, I didn't protest. I could not. It was against the custom for a girl to say no. I knew it was a bad thing, but resistance would have been futile. Perhaps that is why I have been resisting everything ever since – injustice, oppression, exploitation.

'I could not disclose to anyone the shame of living with him. We came to Dhaka. At that time, I was pretty, and he wanted to put me to prostitution. He wanted to force me into this thing, not by selling me to a brothel, but by inviting friends to the house. He used to say to me, 'Why be shy?' He wanted me to have sexual relations with his friends, and then he would get payment from them. It would have been easy. I never told anyone what he was trying to force me to do.

'I left his house. I did not tell my relatives what had happened. They thought I was a fool, because they did not know why I would do such an unheard-of thing – walk out of the house of my husband. I thought it was better to go to the court, to get a magistrate to give me a divorce. This was during the time when Bangladesh was still East Pakistan.

'I developed a hatred for this man. He came to take our child away from me, which he did. Hoodlums came to take me back to him by force. The people who were giving me shelter were threatened. There were five or six terrible years. But I took my son back and kept him with me.

'I had no livelihood, so I started to work in a garment factory when my son was very small. I think the suffering I knew led me to develop a great sympathy for others. Mine was not an isolated incident, but was occurring daily all over Bangladesh. I developed a hatred for society and a determination to change it.'

Chapter Sixteen

If Mayhew had been able to follow more intensively the life of some of his interviewees and had been able to cross the – then – insurmountable barriers of class, he might conceivably have come across the equivalent of Iqbal, who guided me through the labyrinth of child work in Dhaka.

Iqbal is now fifteen. He lives in a slum at Mirpur in north Dhaka and started work as a street vendor of bananas and eggs. But something extraordinary happened to him – something as startling for a boy from the slums of Dhaka as the vicissitudes of the heroes and heroines of nineteenth-century novels who discovered long-lost parents or found out that they were really of aristocratic origin. Iqbal was at school in Mirpur. One day a Dhaka NGO, DRIK, an innovative and original photographic studio, visited the school with the intention of involving the students in presenting, through pictures, their view of the city. Iqbal decided to concentrate on working children. As a result, he was later taken on by DRIK, works for a modest wage and takes as his material the lives of the children he knows.

I wondered how Iqbal would cope with this radical change in his life. Would it confuse him, raise his expectations, wrench him out of the context of his life, separate him from his peers, leave him in a limbo of insecurity? Not at all. Indeed, I could scarcely imagine a child in the West with a more sober sense of himself, with a more mature and thoughtful view of the world. This arises in part out of his strong sense of responsibility towards his family. The labour of children serves as a powerful impetus to growing up and engages them with the world in a way which promotes understanding of the society in which they must survive. The deprivation that his family has suffered has sharpened his perspicuity, enhanced his perception of injustice and the injuries done to him and those like him, whose childhood has been sacrificed for the sake of the survival of those he loves. Iqbal has become the protector of his own parents, their shield against destitution, their best hope of not being stranded in old age. This is a lesson which the children of the West, mercifully, rarely have to learn now, but whether being relieved of such instruction is an advantage to them in their lives may be open to question.

Of course, it is impossible to generalise about the positive elements of child labour. Many children are broken by overwork, used up by their labour, brutalised by ill-treatment. Perhaps the consequences of child work depend upon the sensibility of the particular child. For every one who is strengthened by it and who resolves to overcome the hardship of early years, there must be as many – and perhaps more – who are beaten into acceptance and submit to a lifetime of servitude. Their testimonies should also be kept in mind, even while celebrating the endurance and resolve of a boy like Iqbal.

Iqbal's family – his mother, younger sister and baby brother – live in a one-room hut of *chetai* and bamboo with a polythene roof in Mirpur Two: it seems that the settlements of the poor scarcely merit a name, a number being sufficient to distinguish one from another.

Shahid-ul-Alam of DRIK is an internationally known photographer. He told me how they had selected Iqbal. He said that after grade five, at about the age of eleven or twelve, half the children in Bangladesh drop out of school, because after that time education ceases to be free and many children must make up a portion of the collective wage necessary for survival. For the group of ten children in Mirpur who, it had been agreed, would help DRIK with the project, it was decided that DRIK would help the families financially, so that the children would be able to continue their education, with a little extra also.

But that was not enough for some of them, who were actually the main support of their family. They would find it difficult to continue to work for us, even if we paid the cost of their schooling. One girl, Shapna, was working in a garment factory. We arranged a private tutor for her, but she was leaving home at seven in the morning and reaching home only at ten o'clock at night. She was studying by candlelight until eleven or twelve. The children all desperately wanted to continue their studies. We found a way out for Shapna – we got a loan for her brother so he could set up a small garage. This freed the family from dependency on her income. This was, for her parents, a major sacrifice.

The families of all the children valued education highly, so we had their support. Since the children worked with us, they have changed, become more confident. Their position in the family has also altered – they are no longer so subordinate. We found that our intervention has radically reshaped the pattern of relationships, and that is something we have been acutely aware of. It has disturbed the status quo. Parents see the changes in their lives as creating social and personal problems. The girls in particular have become less submissive, and the parents have had to take a lot of criticism from the community.

Iqbal's father wanted him to start work in a government job. To secure this, he had made a deal with an agent who had promised a job in a government office. He had borrowed a lot of money to pay a huge bribe. He was cheated. There was no job. Iqbal had not wanted it anyway, and this became a new source of friction between them. The money was lost. Iqbal's father was heavily in debt. Then a new baby was born. The whole family was under a great deal of stress.

Shahid-ul-Alam knew nothing of this at the time. Iqbal was not eating properly, but he was too proud to say anything about it. This led to a crisis for Iqbal. He couldn't study properly, he had to earn. 'We got him a job trading in bananas. Banana selling is a big risk, because they spoil very quickly if they are not sold. So Iqbal took a loan to sell eggs. He bought a sewing machine for his mother so she could work from home. She was making carrier bags out of jute – old cement sacks with handles sewn on to them – but these were being displaced by plastic carrier bags.' These do not decay and are polluting all the ponds of Dhaka, multicoloured bubbles of green, pink and blue.

'Selling eggs was all right', Iqbal said, 'until the price of eggs went up, and that wiped out the profit margin. So I had to stop that too.' 'Eventually', Shahid-ul-Alam said, 'we thought it would be the most sensible thing to bring him here, because he needs both an education and the means for his family to survive, and we could accommodate both.'

Iqbal says that if he doesn't work his family will not eat. He would rather study than work if his family could survive without him. Working at DRIK is not so risky as selling eggs. At fifteen, Iqbal, personable, intelligent, eager, nevertheless carries within him, like many children in Bangladesh, an epic of dispossession. His family still know hunger. During the time I knew him, their hut was demolished to make way for a new building. They rented a slightly superior house of corrugated metal, but the rent for this was 1,400 taka a month. Iqbal's income was 2,200 taka; his mother earned a little through the use of her sewing machine. His father was making a pitiful income from collecting the waste from garment factories and selling it.

In the morning, they eat *pantha*, which is rice from the previous day boiled up in water; at midday hot rice and vegetables, in the evening *roti* and dal. Iqbal takes two or three *rotis* from home in a small plastic tiffin can. They never eat fruit, fish only occasionally, rarely meat. Iqbal is himself pitifully thin. He cycles from Mirpur

to Dhanmondi each morning – a distance of fifteen kilometres –
and home at night.

Iqbal was born in Dhaka. Originally the family came from
Madaripur.

Our land was lost through river erosion. My father used to tell
how it happened. My grandfather had his own piece of land which
he used to farm, produce from which he used to trade. He lived
with his brother and my great-grandmother in the house they
had built. The river was eating into the land, but it seemed to be
a long way from the house. My grandfather was always saying,
'We will move eventually, but the river won't come overnight.
We shall have warning enough.' Then in the middle of one night,
one of the two rooms in the house suddenly collapsed: the river
had come and was flowing through the balcony. The land was
lost, together with all the provisions that had been stored in that
room. Everything they had saved was lost.

They didn't come to Dhaka straight away. They looked for
work in the village. But when my grandfather died, it was
impossible for them to stay there. My father came to Dhaka to
look for work. He was seven then. At that age, he had to fend for
himself and send money home. He worked in a factory that made
glue for posters and papers; it was a toxic glue which corroded
his hands. He has been here ever since. He has nothing now.
Since he lost the money he spent in looking for a job for me, it
has upset him so much that he has become unhinged, a little
crazy. But he his still my father. He has done so much for me.

DRIK is a small organisation, but Shahid-ul-Alam says they could
hardly go about their business among the surrounding poverty
without reaching out to the community. The effect on the life of
Iqbal and his family has been profound. Iqbal has taught his mother
to read and write. She wrote a letter of thanks to Alam, which, he
says, is one of his proudest possessions.

When poor children become educated, the parent–child relationship can be easily overturned. Authority is undermined when
children have opportunities which their parents never had. The
intervention of any outside agency creates many changes, the
full effects of which we cannot see, and which can be quite fright-
ening. Yet NGOs go blundering into cultural situations under
the banner of 'education' without the slightest concern for the
consequences. Iqbal says, 'When my sister began to wear
trousers, our landlord objected. I said to him, "When we had
nothing to eat, you never said anything, you didn't object. Yet
when my sister wears trousers, this is important to you. Why? It

doesn't fit?"' On the other hand, when children are educated, they gain new respect from the community. People see we have access to knowledge they don't have. They come to us as counsellors, they want us to write letters, to speak to people in authority. If their children are sick, they ask us what to do. We have learned to take responsibility, and that has never happened before. The children become more adult than the grown-ups.

Iqbal sees his own empowerment as a sign of things to come.

We know that young people in the slums have talents, intelligence and creativity that are all wasted now because their energies are all taken up in the effort to survive. Once your awareness changes, you can never change it back again. Consciousness cannot be reversed, any more than water spilt on the earth can be put back into its pot.

I went with Iqbal through Dhaka, a city in which an ILO survey of 1996 had found children involved in about 300 kinds of economic activity. (In the rural areas the survey found more than ninety.) It showed that 7 million children were working in the whole country, 12 per cent of the working population. About two-thirds work between nine and fourteen hours a day. Ten per cent work up to nineteen hours a day; the average is ten hours.

The self-employed children earned about 800 taka a month ((US$ 20), but in other sectors 363 taka (US$ 9). Forty-eight per cent of them never attended school. Two-thirds of them live with their parents, one in ten live with their employers and 6 per cent live on their own.

The main hazards for children at work were exposure to naked flames, working with electricity, exposure to harmful chemical substances, gas, fumes, garbage, high-speed machinery, sharp equipment, extreme heat or cold, insufficient light, heavy loads, continuous working with ice or water.

Iqbal and I went to Islambagh, a poor long-established district in old Dhaka, an area of winding concrete lanes, small workshops, garages, small businesses and shops. The lanes are deceptive; suddenly, an old wooden door opens out into a courtyard, densely built up with single-room dwellings of concrete or tin; sometimes with flights of stairs to upper storeys, or a ladder leading to a terrace with a splash of crimson bougainvillaea or a palm in a terracotta pot. The next courtyard might offer a long vista of dark concrete corridors, windowless tenement rooms trailing twenty metres or more from the road.

We stopped at a carpenter's shop, a rough concrete space with no window. The only light is from the wooden door which stands open to a lane where cycle-rickshaws, baby-taxis and carts create a perpetual tangle of traffic and noise. A single electric bulb hangs about half a metre from the floor, so that it sheds its faint light on the space where half a dozen workmen sit. They are making small, intricate articles of wood – toggles for coats, wooden buttons – some less than a centimetre in diameter. The concrete floor is covered with dust, wood chips and shavings. The pale crisp furls of wood are collected afterwards and sold at 10 taka a bag as cooking fuel. There are a few metal stools, some chisels, planes and work implements. A man sits at a grinding wheel, sharpening a file which has a fine silver tip. Blue and silver sparks fly as the metal strikes the grindstone.

Ali is employed here. He is eight. He sleeps in the workshop, in the company of one the brothers of the owner. Ali is a small boy, in check *lunghi* and faded shirt. His family lives outside Dhaka. He is clearly unhappy. He works from early morning until late at night, running errands, fetching and carrying, especially observing and learning. As yet he earns no money since this is a kind of apprenticeship: present poverty will be redeemed when he becomes, in turn, a craftsman. That he is working here because this represents a lessening of the burden of expense on his family is not easy for him to understand. Many children are separated from their parents for the same reason that makes them have so many children – they become a cost before they become an asset. It is not that Ali's employer is unkind. It is simply that he cannot offer the tenderness and affection of a family.

The report of the Children's Employment Commission of 1843 in Britain also described 'the objectionable system of apprenticeship'.

> Children, who were frequently orphans apprenticed by boards of guardians, or the children of very poor parents, were legally bound, as early as seven years old, to serve until they were twenty-one. There was frequently no skill to be acquired; often they made one part of an article over and over again, at the end of their term being incapable of making any complete article of what was supposed to be their trade. They suffered often great hardship and great ill-usage, and frequently received no wages for their labour, but only food and clothing, whose quantity and quality varied considerably. Yet to leave such employment meant gaol, for they were held to be legally indentured. (Gregg 1990:136)

The practice is the same, the principal difference being that in Dhaka bureaucratic control of the children, who caused such anxiety to the authorities in Britain, is absent. They are free to depart and to wander the streets, to sleep on the footpaths, to disappear in the formless crowds. But the principle is the same – the promise of skills, only grudgingly imparted, binds children to employers sometimes for years on end. They are unable to judge whether or not they will be supplied with any marketable skills for the future.

When we went back to Islambagh six weeks later, Ali had gone. 'Oh, he went because his parents wanted to send him to school' was the official version. In his place was Chipon, who is ten. This is his first day at work. His father is in a plastics factory, his mother in garments. There are three sisters, but he is the only boy. Chipon is a different kind of child. He goes to school in the morning, works from midday until late and lives with his parents. He does not exude the same feeling of distress and oppression as his predecessor.

We met Shamin in the market at Islambagh. He is a bandhu servant in the house of a small trader. From six to eleven in the morning he works in the house. The *malik* is a milk seller, so for the rest of the day, Shamin goes around the streets selling milk. He does all the errands for the family, shopping in the market. In the evening, he returns to the house and works there until ten or eleven at night.

Shamin was a frightened waif; he was wearing a brown shirt, shorts and green plastic chappals. His mother, he said, was in Barisal, but he had no relative in Dhaka. Sometimes, his employer gave him 50 taka to send to his village, but he had no earnings of his own. We offered him 20 taka, but he shrank from it, as though frightened his employer would accuse him of stealing it. Shamin had no holiday, no free time. He had never been to school, and could not read or write his own name. When we met him, he was on his way to buy vegetables. He was uneasy while we were speaking to him; Iqbal said he would also have to account for his time when he returned. At last, Shamin took the 20-taka note and his scared eyes met mine for a second, a flicker of recognition that we wish him well.

Iqbal said the worst families are not always the rich, but those who can just afford to employ a servant and who begrudge every minute they do not spend in their service, even when they are not paid. Such employers feel that to give board and lodging is itself enough and that the child should be grateful. Shamin had no protector, no one to see whether he is treated unjustly, abused, beaten, overworked. He hurried away into the crowd, his green sandals chattering on the stones, a little plastic bag of spinach and aubergines clutched in his hand, together with the crumpled 20-taka note.

Rubel was working in a small company which makes dies and presses for moulding plastic sandals. In Islambagh, there were many plastics factories – three- or four-storey buildings with machines packed as closely as possible. Rubel was ten. He was earning 10 taka a day (20 US cents) and worked seven days a week. His father was a migrant worker in Malaysia on a rubber plantation. He had two brothers and three sisters. One brother was working as a welder. Rubel's home was nearby. He was learning to operate a drilling machine and worked from nine in the morning until ten at night. He started work two months earlier. His shirt and jeans were greasy. He said he was happy to be working because this would give him a skill which would one day guarantee him good money. Serious, frowning with concentration, he returned to the wheel at the side of the machine, which pierced thick rounds of metal; as the drill spun, it created long silver threads of swarf, which would later be collected and resold. Rubel did not go to school. He said, 'What am I doing if not learning here? This is my school.'

We passed through a low wooden door set in a concrete wall and found ourselves in a rough stony courtyard. All around the yard were single rented rooms, built mainly of wood, with corrugated metal roofs, burning hot beneath a searing May sun. In the very first room a man was making cardboard boxes. Half a dozen children were seated, cross-legged, on the floor, a caricature of a schoolroom. Beside them, piles of cardboard rectangles of different sizes. Dexterously, they took a piece from each pile and, with a few deft twists, turned them into boxes for sweets, shoes and other small items of consumption. The finished boxes were piling up, filling half the room. The younger children, eight or nine, children earned 20 taka a day, the older ones 25. The 'employer' was a subcontractor to a company that supplies boxes and packaging to factories.

The houses in the yard had obviously been built some time earlier; many had been repaired with plastic or cardboard. They belonged to the owner of the factory where plastic shoes were made. The rough ground, the wood-and-tin shacks were reminiscent of a set for a Western film – improvised shacks. Iqbal's uncle was living here with his wife and two children. Inside the hut, there was a large wooden bed and a fridge, which effectively filled the entire room. There was an electric fan and one light bulb. Sharif was foreman in the shoe factory, earning 4,000 taka a month.

A neighbouring courtyard we entered was even more crowded. Hens and goats scattered as we approached. These houses were mostly of metal, stiflingly hot. Each one was home to six or seven people. The furnishing was the same in almost every one: a large wooden bed with a bedroll, a string of clothing, a cupboard containing vessels and utensils, bags hanging from nails in the wall which keep food out of reach of predators. Nurislam was fourteen and working in the shoe factory. His working day was from eight

in the morning until ten at night. He did not go to school and had
already been working for three years. He was operating a machine
that moulds sandals and was earning 200 taka a week. He was paid
at piece rate, 3 taka for a dozen pairs of chappals. He was the oldest
in his family, and his father was a cycle-rickshaw driver.

Monir, twenty, had been in the factory for four years. He was
working from eight until ten, sometimes eleven. He had left school
after class five, although he had worked part time while he studied.
He was now earning 250 taka a week. There were about one
hundred workers in the factory. It was cramped, hot and dusty.
The workers were always suffering from chest and respiratory
ailments. Monir said he didn't think about the future. Survival is
always in the present tense. The work of children is a response to
immediate need. Education, said Iqbal, is a promise for the future,
but the future is a luxury for the poor, because if they don't earn,
they'll never get there. What Iqbal didn't know is that Dhaka is
full of young people who have had an education but remain jobless.
The correlation between education and work is highly unpre-
dictable: one is no guarantee of the other. To those denied
education by poverty, education represents dream and hope: the
capacity of the economic structure to fulfil those dreams and hopes
is another matter.

The rent for the houses in this row is 1,100 or 1,200 taka a
month. Since work is available locally, the rents are high. In the
yard, there is one water tap. Children give their wages to their
parents. The women stand on the threshold of their house, curious
about who has come. Everywhere children press, peering between
the legs of the adults. It is a scene from the crowded alleys of a
Victorian city in Britain, even though there the houses would have
been back to back, and cold and damp rather than burning beneath
the summer sun. But these are details – the sameness of work and
want is stronger than all other circumstances.

In this neighbourhood, we saw many tanneries where hides were
being treated for leather, which is now, after garments and shrimps,
the third largest export of Bangladesh. The small tanneries were
open to the street. Cart-pullers were drawing piles of newly skinned
hides, still leaking their fat and blood on to the stones, and
surrounded by swarms of iridescent flies. The overwhelming stench
of the skins as they were wheeled into the tanneries filled the air
with a nauseating reminder of animal decay which is more repelling
than the decay of vegetable matter. The waste and ordure were
washed away into the public drains, which here and there
overflowed into stagnant pools.

Islambagh is also a major centre for recycling – go-downs, huts
and buildings are crammed to the door with broken glass, a
dazzling accumulation of deadly crystal in the sunlight; stores of

plastic, cardboard, metal, to which children bring their daily scavengings. Islambagh has the appearance of an organised exercise in the careful husbanding of every precious piece of material, from the bones of animals to the smallest splinter of broken glass. Only the bodies of the people are wasted, in every sense; a vast conserving of material things, apart from the poor perishing flesh and blood of humanity.

Later, we visited the plastic sandals factory, where Iqbal's uncle is foreman. This is a three-storey building which looks as though it was originally intended for residential purposes because it consists of a number of small rooms on each floor. The doorways are low, the chipped concrete steps flanked by metal railings. Each room accommodates only one or two plastic moulding machines. These are operated semi-manually. At the top is a square metal funnel, into which the coloured plastic granules are poured. These are melted and compressed into a mould that is inserted manually at the base of the machine. Each mould is placed in the machine singly, and as the operative pulls a lever the plastic fills the space inside. The mould is then opened and the hot plastic is removed: the shape of the sole of a sandal, with an overflow of cooling plastic, which will later be pared by hand.

This is essentially the work that will be done by children. The moulds slide across a greased surface so that they may be set in place faster by the operatives. The finished shoe is thrown on to a growing pile of sandals. The overflow from the top of the mould is torn out by hand and thrown into a waste bin for recycling, together with any defective sandals. The room is oppressive and dark, with a single light bulb. Vats of plastic granules create dazzling splashes of colour – orange, lime green, rose, vermilion. The door of this room is a thin plywood partition, frayed to splinters at the base. The room was colourwashed blue, but it is now dingy, hung with dusty cobwebs, and dust lodging on the uneven surfaces of the walls.

Upstairs, a small room with a pile of green sandals about half a metre high in one corner, orange sandals in another, waiting to be trimmed. There is also a machine which will impress a ridged pattern around the finished sole and another for making the holes through which the toe-straps of extruded plastic of matching colour will be fitted. The boys work with sharp serrated knives to pare away the unwanted material. It is a slow, labour-intensive work; a twelve-hour day for eleven-, ten- and nine-year-olds.

Nanu is eleven. The youngest of three children, he has worked here for three years. An older brother works in a shop. His father sells vegetables, his mother works in a factory making plastic hairbands. Nanu was born in a village in Shariatpur, where the family has a homestead and his grandmother lives. Nanu earns

800 taka a month (US$ 16). His thumb is criss-crossed with cuts from the knife. He gives all his earnings to his parents. His ambition is to become a furniture *mistri*, a carpenter. He would prefer to go to school. He was happy there and felt sad when his father took him out to send him to the factory. The education of some working children has been interrupted for reasons other than family poverty: Nanu ceased school at eight, when three members of his family were already working. He does not voice the question that might challenge parental authority and wisdom.

Nurislam is now fourteen and lives in one of a row of tin huts close to the factory. He has worked half his life, but spent five years in another factory before he began to make plastic chappals. He earns 1,200 a month. (US$ 24). His family came from Vikrampur before he was born, forced out by river erosion. Although he has experience only of urban life, his is a pre-industrial sensibility, formed by a memory of paddy fields and lost lands, bodh trees and wild mangoes there for the eating. He goes to work at six in the morning and returns between eight and nine at night. There is a one-hour meal break, but no food or refreshment is given by the factory. Nurislam studied to class three. He has two brothers who also work in the factory. He dislikes the work that has consumed half his life, the heat, the excessive hours of labour, the chemical pollution. He makes about 500 shoes a day.

Ali Akbar is six, the youngest worker in the factory. He is a helper to the plastic trimmers and earns 400 taka a month (US$ 8). He works a thirteen-hour day and has been working for three months. A tiny withdrawn child with close-cropped hair, he seems traumatised by the factory; bemused, unsmiling, one of the most cruel victims of labour; he could not tell us of his family circumstances.

It is Friday, and the children have a day off. To find some of the other workers, we crossed the Buriganga river from Islambagh to Koylaghat. Through a narrowing street which ends in a pile of half-submerged sandbags and squelchy garbage, to a reinforced sloping concrete embankment, and the expanse of about 250 metres of the fast-running silt-grey river; small shallow country boats worked by one man with a single paddle serve as ferry-taxis. On the other side, the boat arrives at a small bamboo jetty, beneath structures on bamboo stilts planted in the river bed. The jetty itself is a half-dozen poles tied together and cradled by crossed poles driven into the river bed. In the water, all the rubbish has been driven to the edges, green coconut shells, black plastic bags, uprooted water hyacinths that float in the turbid water.

It looks as though the whole settlement has risen out of the river itself. More sandbags and a causeway of bricks in the swamp, into a crooked maze of streets about one metre wide, brick, tin, *chetai* houses in densely packed rows. Rough slabs cover the drains which

carry filth and waste to the river in foul indigo streams that leave
a dark bluish stain on the stones. The river is crowded with boats
which rock and bob, narrowly missing the children who are
bathing, swimming or searching for anything marketable in the
shallows – Mayhew's mudlarks.

We went to the house of Shaiful, a one-room tin building, with
a low doorway. Inside, a large wooden bed, a ceiling fan with a
rusting protective mesh of metal; a cupboard, a shelf of cooking
vessels, a string of clothes. This area is called Muslimbagh. Shaiful
is out playing on his precious day of freedom. Shajida, his mother,
invites us in. Shajida's husband is sick, mentally ill. The family
came to Dhaka three years ago from Faridpur, leaving their two or
three decimals of land around the homestead in order to seek
treatment for the sick man. There are four children, three boys
and a girl. Shajida's husband works intermittently, selling cups
and plates, but he earns very little. Two boys are working in the
sandal factory, and the girl, a beautiful fourteen-year-old called
Sweetie, is in a factory that makes Muslim *topis*, the lacy circular
caps worn by the devout.

Shaiful is eleven and has been working for two years. He earns
700 taka a month. Together, the three children earn about 2,000
a month, which is just about enough to sustain the family. Although
they are never hungry, their diet is not very nutritious – no meat,
no fruit, no eggs, no milk. Shajida says it would cost 4,000 a month
to eat well; when she has paid the 600-taka rent, they have about
1,500 taka to spend on food. She would prefer them to go to
school, but in view of her husband's infirmity they have no choice.
Shaiful has heard that someone has come to his house. He is a
small smiling child with a dream of going to school. For a moment
he looks at me, and I can see the hope in his face that I might have
come to deliver him from his labour; the faintest flicker of disap-
pointment, but the smile doesn't falter. He works twelve or thirteen
hours a day and cuts 100 shoes a day. In the morning he eats some
pantha before leaving home. For his midday meal he takes tiffin
in a little metal can, some vegetable and *rotis*.

Rubel, Shaiful's friend, is with him. He is nine and has been
working for two years. His job is to clean the machines. His mother
works making marriage garlands, baskets of roses and sweet-
smelling *rajni sughanda* flowers, and his brother works with her.
His father is a construction worker. Rubel never went to school
but he can write his name. He was born in Dhaka; his family came
from Faridpur when they sold their little land. Landless people
who come to Dhaka suffer terrible insecurity: perhaps this is why
they set their children to work, in an effort to recover the mea-
sureless loss of self-reliance; poor waifs who pay the price of a social
dislocation they did not choose and cannot understand.

Chapter Seventeen

One day I went with Iqbal to Kamalapur Railway Station. This is a new building, clean and spacious, its architecture halfway between a mosque and a museum. The central concourse is paved, and an undulating roof rests upon thick pillars. On the pavement lies a bier, with, just visible, the waxen feet of a body, around which the relatives sit, waiting for a train that will bear the deceased back to his village.

On and around the station live dozens of children, ragged, homeless, abandoned, many of them irrepressibly cheerful. Today is a public holiday, *Pahile Boishak*, the first day of the Bengali New Year and one of the major festivals. Families stroll through the park to see the celebrations, traditional dance, song and drama. Children carry balloons and toys, and the park is crammed with stalls of vendors of snacks, drinks, novelties, sweets, fruit.

All this only emphasises the desolation of the children; their half-naked bodies, callused bare feet, dusty skin; some obviously affected by ringworm and scabies, some more clearly ill-nourished than others. Only the ubiquitous smile is present, an almost abstract thing, the sole sad patrimony of the poor of Bangladesh, priceless but without market value.

We went onto the station platforms. Iqbal led me to the station master's office to ask permission to speak to some of the children working on the platforms. The officials seem astonished by our request: they find it incomprehensible that anybody should want to talk to them and, even more so, that anyone on such a bizarre errand would seek permission of those who have no responsibility for them. These children, explain the officials, belong to no one; itself a significant comment, since most children are regarded as an extension of their parents. Not chattels exactly, but without distinct identity apart from the families that define them. This is why people in South Asia usually seek to integrate friends – and even strangers – into some closer familial bond. 'Uncle', 'Auntie' and 'Brother' are common forms of address.

In his discussion of street children in Central and South America, Duncan Green speaks of the emergence of a 'street children industry'. The idea of street children – semi-feral, self-reliant, living in collective groups beyond reach or control of the mainstream – strikes terror into the people of the rich West. They are seen as dirty and dangerous. They have been the object of an almost obsessive NGO concern, who also sometimes see in them heroic rebels, 'urban pirates', who have chosen freedom. Both stereotypes are false: they may have fled violence in the home, only to encounter it again on the streets. On the other hand, there are networks of support, there are comforts and satisfactions to be found in each other's company. There are drugs, addiction to glue, crack and dope, there is sex. But there are vigilantes, death squads and HIV and sexually transmitted diseases.

The drama of Latin America is not much in evidence in Bangladesh. What strikes the observer here is the submissiveness of the children, their unthreatening manner. The violence that corrodes them is more stealthy, since they react so faintly to the ravages it works on their thin bodies. They are more like the wraiths interviewed by Mayhew, those who became the object of 'rescue' by Dr Barnardo, than the bands of menacing street children under the concrete flyovers and in the favelas of Brazil, whose presence is so menacing to the middle class that they extend a tacit complicity to those ready to exterminate them.

Rabiul is eleven. He works as a coolie on the station. He says that at home there was no eating, so he simply ran away. He sleeps on the floor of the station. The police sometimes beat the children and chase them away. They do not extort money from these children since the pickings are so scanty. The few taka they earn or beg daily is spent on food. There are small food stalls lining the traffic lanes in front of the station – bowls of rice, a little dal, some vegetables, potatoes and *brinjal*, some bananas or parts of bananas that have rotted and cannot be sold elsewhere. Rabiul has three brothers and two sisters. His father died. His landless family is in Khulna. He says he earns about 30 taka a day (60 US cents). Of this, he spends 20 to 25 taka on food. He works from six to eight in the evening, but studies at a school run by an NGO 'sometimes'. He has not yet learned to read or write.

Most of the schools set up by aid agencies and charities in Dhaka try to adapt to the necessities of the labour that their students perform. They recognise that learning must be adapted to such times and opportunities as the children may take from work. This is not the only difficulty in organising the children here or at the river launch terminal. For most of these, the family relationships have been broken by death, separation, poverty, abandonment. Many are disturbed and restless, mobile, not motivated to learn. In any case, the basic learning they can acquire scarcely equips them for very much beyond the roughest of manual labour – many will become rickshaw drivers, construction workers, cart-pullers at best.

Billal is ten. He came, he says, from Faridpur, 'as a small boy', although he is small enough now. His mother is sick, and there is no regular food. He sometimes gives money to his family from the 50 taka he earns daily (50 taka sounds a lot. I'm not sure Billal knew how much this was). There are four brothers and one sister. His father also works in Dhaka, as a cycle-rickshaw driver; he lives in Gopibagh, but Billal prefers to sleep at the station. He says that all the children here are his friends.

It is clear that some of the children are here from choice – if running away from hunger or abusive families can be called choice. They present themselves as already adult, heroic, independent. And, indeed, there has developed a culture of survival, a spontaneous protective network among the children, and powerful affective bonds which hold them together; hierarchical, no doubt, with the older and stronger looking after the younger ones, but in some ways probably more satisfying than those in homes blighted by misery and insufficiency. Billal says that when he grows up, he wants to do a regular job, any job. Here, he carries bags for passengers to and from the train. Sometimes they give him 5 taka, 2 taka. He can read and write a little. He thinks he will perhaps become a rickshaw driver when he is old enough. That is the one job that is open to any poor man who is reasonably healthy since it requires neither training nor investment, apart from the cost of the hire of the vehicle (about 40 taka a day, 80 cents).

These voices are those of the direct descendants of Mayhew's girl crossing-sweeper with its sweet resignation.

I'll be fourteen, sir, a fortnight before next Christmas. Father came over from Ireland, and was a bricklayer. He had pains in his limbs, and wasn't strong enough, so he gave it over. He's dead now – been dead a long time, sir … I wasn't above eleven at the time. I lived with my mother after father died. She used to sell things in the street. Yes sir, she was a coster. About a twelvemonth after father's death, mother was taken bad with the cholera and she died. I then went along with grandmother and grandfather who was a porter in Newgate market. I stopped there until I got a place as servant of all work. I left them because they was going to a place called Italy. I went back to grandmother's, but after grandfather died, she couldn't keep me, so I went out begging. I carried lucifer-matches and stay-laces at first. At last, finding I didn't get much at begging, I thought I'd go crossing-sweeping. I saw other children doing it. I says to myself, 'I'll buy a broom.'

Imran, twelve, from Chatpur, has been here a year and a half. He left home because his father had taken a second wife and this stepmother treated him cruelly. He has two brothers and two sisters. He works from six in the morning until late at night, and then sleeps on the station floor. He earns between 40 and 50 taka a day. Imran spends 30 taka a day on food and keeps clean by using the station washroom, while a nearby restaurant provides him with drinking water. He pays a few taka daily for this. He has no money to send home to his family. He does not like the station, but cannot go back because he is not wanted there. Imran does not want to be a rickshaw driver when he grows up. He is sometimes frightened by *mastaans*, thugs, some of whom have guns; although, he says, they do not bother the children.

Furkan, who is twelve, is from Chittagong. He came to Dhaka because his father died and his mother remarried. His second father is not good. The family is not poor, he says: they own two houses. There are two brothers and one sister. Furkan earns 30 to 40 taka a day. Here, life is okay because he has friends. He goes to school 'sometimes', but cannot read or write. He would like to go back to his village to become a farmer. His parents know where he is living. They do not care. He works from early morning until late in the night. Sometimes people give him 1 or 2 taka for carrying their bags. Occasionally, they beat him when he asks for money.

Bilkis Bhanu is a girl of about eight. She does not know her age. She carries her baby brother, about eighteen months old, in her thin arms. She has ragged matted hair and wears a pair of pink trousers, but the baby is naked. She lives with her family on the road and survives by begging. Her father is sick with cancer. She is lucky if she makes 10 taka a day. Today, she has so far earned 2 taka. She shows the two coins clenched in her hand. She is the only girl in the family, and there are two younger brothers. Some days they eat once, sometimes twice; occasionally not at all. They came from Kishorganj, outside of Dhaka, to find treatment for her father's sickness. The hospital gives free medicine. She has never been to school and wants to work because working will earn her more money than begging.

Mayhew had much to say of beggars in mid-nineteenth-century London. He was especially stern against those who pretended some lawful occupation, the better to excite the compassion of passers-by. He singled out those he identified as the 'petty trading beggars' for special scorn.

This is perhaps the most numerous class of beggars in London. Their trading is in such articles as lucifers, boot-laces, cabbage-nets, tapes, cottons, shirt-buttons and the like, is in most cases a mere 'blind' to evade the law applying to mendicants and vagrants. There are very few of the street vendors of such petty articles as lucifers and shirt-buttons who can make a living from the profits of their trade. Indeed, they do not calculate upon doing so. The box of matches or the little deal box of cottons is used simply as a passport to the resorts of the charitable. The police are obliged to respect the trader, though they know very well that under the disguise of the itinerant merchant there lurks a beggar.

In Bangladesh there are, of course, no laws against beggars; but they are no less at risk from the police, who may extort money from them for leaving them unmolested. Nor is there active in Dhaka any organisation such as the Society for the Suppression of Mendicity which Mayhew congratulated for having cleared the streets of nearly all impostors in recent years.

He is no more charitable towards children.

The lucifer droppers are impostors to a man – to a boy – to a girl. It is children's work, and the artful way in which boys and girls pursue it, shows how systematically the seeds of mendicity and crime are implanted in the hearts of the young Arab tribes [sic] of London. The artfulness of boys is of the most diabolical kind; for it trades, not alone upon deception, but upon exciting sympathy with the guilty at the expense of the innocent. A boy or girl takes up a position on the pavement of a busy street. He or she – it is generally a girl – carries a box or two of lucifer matches, which she offers for sale. In passing to and fro, she artfully contrives to get in the way of some gentleman who is hurrying along. He knocks against her and upsets the matches which fall in the mud. The girl immediately begins to cry and howl. The bystanders, who are ignorant of the trick, exclaim in indignation against the gentleman who has caused the poor girl such a serious loss, and the result is either the gentleman, to escape being hooted, or the ignorant passers-by, in false compassion, give the girl money.

His response to the destitute, and to those who have recourse to any expedient to get money, is echoed a thousand times by the middle class in Dhaka and the stories they tell one another in their secure apartments in Gulshan and Banani. Sometimes, it seems they must have read the pages of Mayhew, as they solemnly narrate identical stories. Sometimes their fables are even more lurid, of parents who amputate the limbs of their children in order to make them more proficient beggars, those who mutilate their young so that they might have a good start in life! They tell each other that the people on the streets are not really in want; that they are organised into chains of beggars who live like kings on their daily takings. And all this is said in the presence and in defiance of visible consumption, malnourishment and sickness.

It is, of course, part of the defence mechanism of those who must believe that the legions of the absolute poor form part of a conspiracy to induce them to part with the wealth which, they are convinced, they have worked for by the sweat of their brow, which they have merited by their intelligence and acumen. These qualities, of course, when applied by the destitute, become cunning, dishonesty and a capacity for swindling; maybe precisely because the well-off see themselves mirrored in the outcasts of the streets. If this is the case, they must of course distance themselves from such unwelcome kinship, simply in order to go on with their lives. Who could live with the thought that the child of about ten, rocking to and fro over the body of his dying mother close to the High Court in Dhaka, could be anything but a fraud? Who otherwise could step over the scenes of existential desolation which would stop the world if it paused to reflect on its causes?

Mayhew again evokes sights familiar on contemporary city streets when he speaks of cripples:

> The poor wretch without hands, who crouches on the pavement and writes with the stumps of his arms; the crab-like man without legs, who sits strapped to a board, and walks upon his hands; the legless man who propels himself in a little carriage constructed on the velocipede principle ... I cannot think that the police exercise a wise discretion in permitting some of the more hideous of these beggars to infest the streets. Instances are on record of nervous females having been seriously frightened, and even injured, by seeing men without legs or arms crawling at their feet.

Mayhew obligingly spells out his prejudice and that of those he felt bound to protect even from the sights with which their indifference had strewn the streets of London. At least the equivalents of the 'nervous females' in Dhaka would probably be less troubled by scenes of even greater dereliction, towards a daily contemplation of which they exhibit a well-practised fortitude.

The dingy station children are part of the landscape. No one looks at them. Even those who employ them to carry their bags scarcely give them a passing glance, but hold out a 2-taka note at arm's length. The children have no one but each other; epithelial marks of cuts and knocks from heavy cases, cropped hair, discoloured bits and pieces of garments; teeth silver in dusty skin, mineral eyes of copper or bronze. No wonder no one looks into them; it is too painful.

We move on to the platform. A concrete seat girding one of the massive pillars. Soon a crowd collects, so that talking to the children rapidly becomes a piece of street theatre. Kokon is twelve. He wears a striped shirt and a rough cloth around his shoulders. A bandage covers the lower part of his leg. Two days earlier, he had an accident when a fish barrel fell on him as he was unloading it from the train. He came to Dhaka a year ago and is alone here. He has no brothers or sisters. His parents are both dead. He sleeps on the station. Today he has so far earned nothing (it is noon). He expects 15 or 20 taka a day and eats in the stalls outside the station, sometimes twice, sometimes three times. In the morning he pays 3 taka for *roti* and tea. In the evening rice and vegetables costs 5 taka. Sometimes the police come and beat the sleeping children with their *lathis*. When he is sick, Kokon sleeps throughout the day as well. He has no relatives. His father fell sick and died, and his mother died in childbirth soon after. Kokon has no clothes other than those he wears. He washes himself in the river.

Alamgir is seven. He works and sleeps on the station. He is from Sylhet and has been here one month. He was brought to Dhaka by a stranger and abandoned on the station. He has two brothers and four sisters at home. He gets a little money by begging, 10 or 15 taka a day. He eats rice or *roti* with vegetables, but on days when he makes no money, he will eat rotten food and fruits that have been thrown away. He wears a pair of short pants, dark grey, with an elasticated top. This is his only possession. No shoes, no shirt. He is as bereft of belongings as it is possible for a human being to be. The existence of this skinny starveling surviving on less than 20 US cents a day is a reproach to a world and the glory of its wealth-creating capacity, a market which cannot reach him with its delicate sensory equipment that can detect only money.

Monsur thinks he may be twelve. He earns only by begging, and if he is persistent he might make 20 taka a day. Some days he will get five or less. He carries the bags of travellers. Sometimes they pay him, sometimes not. If the bag is very heavy, they may give him 10 taka. He has come from Mymensingh. His father died and when his mother remarried, the new husband did not want him around. He came to Dhaka when he was seven and has lived on or around the station ever since. He does not know any other part of Dhaka. He is alone and does not know where his brothers and sisters are. When he grows up, he wants to own a cigarette shop. He also has no clothes other than the shirt and pants he wears. He washes them only in the rain. When he is sick, if he has fever, he just lies on the station floor until he feels better.

We went to the New Market, a place which serves as terminus and starting point for the Tempo taxis. These are vans, with two rows of wooden seats facing each other, which can hold about five people each. The helpers – who collect the fare and yell out the destination of the vehicle for travellers – are often children; the smaller the better, since they can cling with their hands on to the roof of the Tempo, with their feet balanced on a metal bar at the back of the vehicle. If they are very small, they can move quickly in and out of the interior. They stand precariously, small-denomination notes folded between the fingers of their left hand, a pocket of their ragged jeans full of coins. As the vehicle lurches through the traffic, they sometimes lose their footing and are thrown on the road into the path of following traffic. Some are barefoot, others cling on with prehensile plastic chappals. When the passengers are ready, they whistle to the driver as a signal to set off. They compete with others for travellers; they are jaunty, knowing, exchange jokes with passers-by. They are soiled by fumes, oil, dust, street dirt. The work is arduous, hazardous, extremely unhealthy, for the air of Dhaka is among the worst in the world. Many will themselves become drivers when they grow up.

Jakir is twelve and has been working as a helper for a year. He lives with his sister and her husband not far from the New Market. Three other sisters and two brothers are at home with their parents in Faridpur. Jakir earns 60 taka daily, but he must start at six in the morning and work sometimes until twelve at night. He eats meals quickly, snatched at a stall close to the Tempo stand, and this is where we caught him, ravenously consuming rice, *roti* and vegetables for which he had paid 8 taka. Jakir can read and write. His family have a very small piece of land, but not enough for survival. He likes Dhaka because here he can find work and never goes hungry. Sometimes at home – he grimaces. Somebody had to come to earn some money, he says, and as he speaks you can see the pride in his eyes. 'I didn't mind. I send money home, 300 to 400 taka a month, so I know that my brothers and sisters get enough to eat.' When Jakir grows up, he doesn't want to be a Tempo driver. He wants a good regular job with shorter hours of work. He says he has never had enough sleep since he came to Dhaka. His eyes are ringed with fatigue, his hair is dishevelled, his whole body exudes the sour breath of the city.

Mutlub Ali is eleven and has been a helper for five months. His father and mother live in a village in Dhaka District. He is staying with his brother. At home there are two other brothers and two sisters. Mutlub Ali earns about 50 taka a day and sends 400 to 500 taka a month to his parents. His money is paid by the driver, and the little boy works from seven in the morning until eleven at night. Sometimes he is so tired in the evening that he almost falls asleep. He doesn't like it and wants 'a proper job' when he grows up. This usually means secure employment, preferably in government service.

While he is talking to us, Mutlub Ali does not cease work. He is trying to turn the vehicle around, pushing it by hand so that it is facing the direction in which it will travel. Then, in between answering our questions, he starts to call out for passengers. He has a shrill hoarse voice when he shouts and the voice of a child when he talks to us. There are few intervals from labour; their time, every moment of it, is colonised by work and belongs to whoever employs them. The Tempo is soon full. It departs in a mushroom of black fumes. Mutlub Ali jumps deftly on to the back, and is gone.

Changes in technology scarcely conceal the enduring values that are so easily transferred through time and space. In Britain, chimney sweeps were characteristic figures among child workers. They, like the Tempo helpers, were selected for their smallness, since they were required to climb the narrowest chimneys. Sometimes they were purchased for as much as £5 apiece, occasionally they were kidnapped. In the large towns they worked with their masters for twelve to sixteen hours a day. They began work at six years old, 'a nice trainable age' as one master said.

Chapter Eighteen

Direct parallels between the nineteenth and twentieth centuries may, of course, sometimes be obscured by dramatic changes in technology, so that the echoes and correspondences are obscured by conditions which themselves seem to have transformed the landscapes of industrial society. One group of Bangladeshi children exploited far from home, reminiscent both of the slave children and industrial workers, is nevertheless used in the entertainment industry – the camel jockeys sent to the Gulf for the amusement and profit of the rich. The *Sunday Telegraph* of 30 November 1997 reported that 'Thousands of children are sold or kidnapped from the age of two to five years and sent to Abu Dhabi and Dubai for camel racing. They are starved to make them lighter, and they sleep on the floor of corrugated metal huts. Parents who sell their children usually receive between 12 and 20 pounds.'

The children are usually strapped to the camel with a rope, and the camel is then whipped into a frenzy. It is urged on by the screams of the terrified child. The preferred weight of a camel jockey is nineteen to twenty kilograms, and the upper limit is usually thought to be forty. The younger and lighter the child, the louder he screams, the faster the camel will move. Many children are mutilated or killed when the security rope which is supposed to hold the child fast works loose. Before 1993, when the practice of using child jockeys was banned in the United Arab Emirates, up to twelve children a week were killed.

The Gulf started 'recruiting' children from Pakistan in the mid-1970s. Parents were often deceived into parting with their children with the promise that they would be entering safe and guaranteed work. In the beginning, the trafficking was done by air, but later, as a result of international pressure – and the sheer volume of the trade (some 19,000 children are known to have been trafficked to the UAE) – the sea routes were followed.

It was reported in September 1997 that thirty-eight Bangladeshi children of between three and four years of age were in the care of an NGO in Madras, after being rescued six months earlier while being trafficked to the Gulf as camel jockeys. Some said they had been sold by their parents.

The United Nations Special Rapporteur's interim report to the 1998 General Assembly notes concerns related to the risk to the lives of young boys, some as young as four years old, who are trafficked from countries in South Asia to supply the demand for camel jockeys. 'The children are attached to the camel's backs with cords and those who fall risk being trampled to death by the other camels on the track. If the children refuse to ride, they are beaten.' The Special Rapporteur noted in 1997 that the Camel Jockey Association of the UAE finally prohibited the use of children as jockeys, but cited evidence indicating that the rules were being ignored. In February 1998, ten Bangladeshi boys aged between five and eight were rescued in India after being smuggled as camel jockeys. Also in 1998, airport officials rescued two boys being taken to Dhaka from Sri Lanka by two men later charged with kidnapping. Although since 1993 there has been a prohibition on children under the age of fifteen or weighing less than forty-five kilograms, in September 1998, a boy of five weighing twenty kilograms was taken to hospital after an accident in a camel race.

There is one significant difference between the child workers in the mills and mines, the chimney sweeps of nineteenth-century Britain and the camel jockeys and the young girls trafficked for prostitution in South Asia today. The latter are employed for the amusement of adults in leisure industries. Their suffering is itself a spectacle which can command big money. Whatever the horrors visited upon the pauper children in the past, these were for the most part clandestine, secretive, unseen; shame tended towards their concealment. Profit was to be made from their labour, not from the public show of it. The idea that the whole world is embarked upon a steady path of progress that will take us away from such barbarities must be reconsidered when we contemplate the numbers of children entering, not only known forms of bondage, but unimaginable new kinds of abuse within the new integrated global economy, of which the unfortunate camel jockeys are only one example. Given that there are now more slaves in the world than there were at the time of the emancipation of the serfs in Russia and of the freeing of the slaves in America, the faith that wealth creation will deliver freedom to all humanity is apparently, on the present evidence, a pious and scarcely justified hope.

Chapter Nineteen

On another occasion I went with Iqbal to his home in Mirpur. This is a slum area, huts of bamboo frames and *chetai* walls. This district is now not so far from the expanding centre of Dhaka, and the slums are slowly being displaced by multi-storey apartments. A forest of crude wooden scaffolding, piles of bricks and the temporary barracks of migrant construction workers now stand where Iqbal used to live. His house was demolished a few months earlier. His family have moved to a lower-lying spot, about fifty metres from where they were before. Enterprising owners of the land have rapidly erected huts which can be let at 600 or 700 taka a month.

It was a stormy day. Low indigo clouds hung in great swags on the horizon, and a fresh wind disturbed the palms and grasses remaining in the spaces between the huts. We paused at a shop, surrounded by children selling *tokai*, or waste, to the owner. He was weighing the day's findings and paying them according to the value of what they had gathered. Iqbal told them we wanted to ask them about their work. Within a few seconds we were surrounded by a crowd of children who followed us in a noisy procession to Iqbal's house.

The hut was hard to reach because the morning's rain had left great pools around it, the colour of a spillage of milky tea. The dust and earth had been transformed into a sticky paste that clung to the shoes, forming a heavy platform of mud on the soles. In front of the dark nimbus coming from the south-west, rags of lighter cloud like smoke and, at the far horizon, a narrowing strip of daylight, an electric yellow reminder of a day prematurely faded.

The house was a single room with *chetai* walls and a wooden door. Iqbal had made a partition at one end of the room, which formed a narrow bedroom. In the main room, only a large wooden bed, a cupboard, a folding chair, some shelves against the wall. Very little floor space remained. Iqbal's baby brother, eighteen months old, was hanging in a rope chair suspended from the ceiling, swaying in the centre of the room. The children who had followed us crowded unceremoniously into the house. Iqbal's sister, twelve, and his mother ordered them to wait outside. But the rain suddenly started: at first, great droplets that splashed on the polythene that had been pulled tight over the *chetai* roof. But within a few seconds it was dancing in the puddles and became a torrent, reducing visibility to zero, a wall of water splashing around the hut, causing the whole structure to vibrate and sending a thin spray of drizzle through the tiny gaps in the woven walls. Iqbal's mother gave the children permission to come inside on condition that they behave themselves.

Iqbal's room contained a narrow bed of hard chipboard, with a pillow and chadar as bedcover; a wooden bookshelf held some Bengali novels and stories, and some calendars published by DRIK which displayed some of Iqbal's photographs. He had one English book. It is a paperback called *The Age of Napoleon*, and Iqbal was trying to learn English from it with the help of a Bengali–English dictionary.

The working children, who were Iqbal's friends, told their story. Jamal did not know his age. He had lived in Mirpur for 'many years' and he looked about thirteen. His father was a construction worker and his mother in a garment factory. The family came from Barisal, about ten hours by bus from Dhaka. Jamal had two brothers and one sister. He never went to school regularly, but knew how to write his name. He remembered he went to school 'once', but had to leave because the family were very poor and had no land. He said he collected *tokai* from morning until dark and was earning between 30 and 40 taka a day. He got 1 taka per kilo for paper, 3 taka for iron or tin, 2 taka for glass, 8 taka for plastic. The man to whom the children sold delivered it to the factories. Jamal said he wanted a 'proper job' when he grew up, as long as it wasn't in construction. Jamal didn't enjoy what he was doing. His brother, who was seven, had just started the same work. He said the family ate two or three times a day, occasionally only once a day. His family lived in a hut similar to that of Iqbal, and Jamal thought they were paying 30 to 40 taka daily, so his earnings, he said proudly, were paying the rent.

Many children said they do not see what they do as 'proper' work. This suggests what we well know – that much of it is casual, insecure, ill-paid. A proper job means a factory job; it is only the more ambitious who can dream of government work, which represents real security.

Babu, fifteen, was also collecting waste. He was born in his village in Mymensingh, but came to Dhaka when he was three. His purposeful wanderings took him far afield – to garbage heaps, the areas around factories and industrial units. He worked from early morning until night, and earned 20 to 30 taka a day. Sometimes, he said, the children would make a surprising find – a heavy piece of metal or something of value, that might give them a spectacular rise in income. Such days, he said, we are very happy.

Babu's father was living in the village, but his mother was in Dhaka, not working. He had two brothers and two sisters, but the oldest brother was sick and could not work. Another brother was a helper in a garment factory, earning 600 taka a month.

Babu did not like the idea of entering a garment factory. Six hundred taka a month was less than he was earning now; factory work was monotonous and repetitive, and the hours very long – from eight in the morning until nine at night. Collecting waste gave a freedom to move with friends, to rest when you felt like it. Babu's family expected to eat three times a day, *roti* or rice with vegetables, mainly potatoes and *brinjals*. They never ate fruit and rarely meat; occasionally, a banana. Babu knew the price of daily necessities, 13 taka a kilo for rice, which was not very good quality, 9 taka a kilo for flour. It was Babu's ambition to be a Tempo driver. He had worked as a Tempo helper, but the hours were also very long, from six in the morning until ten at night, and the earnings no more than he was getting in his present job.

Roni, eleven, was a Tempo helper. He was working one day on, one off, in an improvised work share with another boy. The hours were very long – from five in the morning until twelve at night – so daily work was impossible. His earnings varied from 10 to 25 taka. He had one brother and one sister, and his father was driving a cycle-rickshaw, his mother worked in a restaurant. He never went to school. The family came only recently from Barisal, where they have no land apart from the small patch where their homestead stands. Roni started work in the village when he was eight, looking after other people's buffaloes. He said he would rather be in the village, because there you could eat mangoes or jackfruit from the trees. Here, you had to buy in the market and it was very expensive. The family ate rice or bread, with potato, onion and chilli. They were living in a one-room *chetai* house, for which they were paying 350 taka a month. Roni was hoping that his apprenticeship as a Tempo helper would one day lead him to become a driver.

Ripon was nine. He carried a basket, selling sweets – *kotkoti* and *tilarkhaza*. He was working from midday until night, making 20 to 30 taka daily. His father had a small dry-goods shop, selling rice, flour and dal. The family came to Dhaka fifteen years earlier. Ripon was born in the city and attended a school run by BRAC (an NGO) between nine and twelve. He would get up between five and six in the morning, help his mother with housework and chores, then go out selling until it was time for school. After school, he went selling again, and would expect to go to bed at ten.

Hassan, ten, was also selling sweets. It turned out that there was a manufacturer of sweets nearby, and the children worked on commission, being paid according to their daily sales. Hassan was going to school and still earning 30 taka a day. His father was a rickshaw-puller, his mother a domestic worker. He was the oldest of three brothers. His ambition was to become a doctor. The family came from Bogra because they had no land, and Hassan was born in the city. He said that he got up at five in the morning, washed his face, did some reading and writing, and then went off after six selling sweets. He returned to go to school between ten and one. After eating at home, he went back to work from about two thirty until ten o'clock at night. Then he would do homework, reading and writing and go to sleep at about eleven. Sometimes he played football, although there was little time for playing in such a busy life. The family had no television, but Hassan occasionally watched it at a neighbour's house. He also had many duties at home, such as fetching water from a public tap. For this, the family were paying 30 taka a month.

Saiful was eight. Sometimes he did domestic work, occasionally sold sweets, and from time to time collected waste from the garbage bins. He gave the money to his mother for rice and vegetables – anything between 5 and 15 taka a day. Sometimes he was called for domestic work, but this was unreliable. When he was required for this, he would wash the vessels, clean the floor, wash clothes, do the ironing. He would be given 10 taka a day for this. His father was a security guard and his sister a helper both in a garment factory. His mother, he said, was begging. Although the family came from Barisal, Saiful was born in Dhaka. He said his ambition was to be helper to a truck driver, carrying goods and travelling all over Bangladesh.

It is perhaps not surprising that children know precisely how their earnings are spent. Saiful said it went on rice and vegetables, others said they helped with the rent, some said they helped their brothers and sisters to go to school. Their contribution is recognised by their parents and gives them a sense of responsibility, indeed, indispensability for the survival of the family. It is assumed that a 'good' son remits his entire earnings to his family until he is about fourteen or fifteen. After that, he may be allowed to keep some money for himself, which gives him a measure of independence. This is sometimes unsettling for parents, because they find it difficult to accept that their interests and the interests of their children are different.

This is also very much in evidence among the young women who work in the garment factories. The single largest industry which has grown in the shadow of the even minuscule amounts of money the young women keep for themselves is cosmetics; which is regarded by many as a sign that factory work is making them immoral, undermining the values of religion and family. It is a paradox that the very thing which makes children work – their contribution to the family – also, paradoxically, creates small spaces for them to assert their own autonomy, however faintly in comparison with Western children.

All the time we were talking with Iqbal's friends, the rain was falling. On this particular day in May 1999, Dhaka had 145 millimetres of rain (about six inches). The ochre-coloured water had risen higher but had not entered the hut, which was built on a small platform; but water had darkened the *chetai* and soaked the brick of the neighbouring buildings blood-red. A late shaft of sunlight illuminated the scene with unusual clarity since all the dust had been washed from the air: a white-painted block of flats was stained persimmon colour by the dying sun; the little bamboo houses were orange and brown, the grasses and palms a new dazzling green, whispering in the cool humid wind. Some rickshaws had been stranded to the top of the wheels, so that they looked as though rooted in water and their perfect reflection was held in the still pools.

By keeping close to the houses, which were all slightly raised, we were able to reach the little brick road. Water was cascading down the uneven surface, gurgling in the drains and singing in the overflowing ditches, turning the heaps of garbage into a black sludge, water dripping and trickling, overflowing in this wet incontinent city.

Many people had come out of doors in the late sunlight, which also intensified the colour of their clothing, scarlet and orange sarees, cobalt and lime green; faces, too, burnished and gilded by the thick sunlight, like the chorus from some epic opera production, as they stepped on the bricks that rose above the water level; it was a powerful mixture of beauty and misery – the very essence of Bangladesh, where natural and human-made disasters merge with a naturally human resilience and splendour.

Here we came across Shuma who is eight. She was wearing a violet-coloured dress and carrying a pitcher of water; a smile a tiny crescent of silver. She is a domestic worker, but had been given a day's holiday to visit her family. She said she lives in her employer's house in another part of Mirpur, but was released from work at two o'clock so that she could go to her mother's house. Shuma said that her employers are very good. She gets up at seven o'clock, washes the floor, cooks breakfast and washes the vessels. She cuts vegetables for lunch, cooks rice, serves the meal and clears it away. Then she prepares supper. She finishes her work at about ten and goes to sleep at ten thirty. She receives no money for her work. She has no father and her mother also works as a maidservant. She has one sister. She was very happy to spend some time with them, because she did not know when she would next have the opportunity to do so again.

Iqbal insisted he would come back with me to Shantinagar where I was staying. First, he took me to the house of a friend, a girl of thirteen called Shamin. We could not go inside her house because the water there was knee-deep. Shamin was a serious young woman, self-possessed, still at school. Iqbal asked her if she would like to come with us in the baby-taxi for a ride.

The air was still now, humid and very cool. The three-wheeler splashed through the flooded streets, sending silver fans of water over the legs of passers-by. We went past the Sheraton Hotel, which was unreachable by the cars and limousines at the entrance because the drains were blocked and the roads flooded. It felt very odd. Today was my sixtieth birthday, and these two children had come back with me as a courtesy and protection – from what? We went into the hotel and ordered mango juice. They became very subdued because, although it was a modest hotel, they rarely enter such places. The staff watched us curiously. I went with them to the gate. I felt confused – unaccustomed to the politeness and consideration of children; another poignant piece of instruction from people whom we usually consider to be under our permanent tutelage.

In his London, Mayhew had observed grimly that 'almost the only commodities in which a legitimate trade is carried on by the petty traders of the streets are flowers, songs, knives, combs, braces, purses and portmonnaies'. He qualifies this by stating that

> The vendors of flowers and songs, although they really make an effort to sell their goods, and often realise a tolerable profit, are nevertheless beggars, and trust to increase their earnings by obtaining money without giving an equivalent. A great many children are sent out by their parents to sell flowers during the summer and autumn. The sellers of flowers are chiefly young boys and girls. They buy their flowers in Covent Garden, when the refuse of the market is cleared out, and make them up into small bouquets which they sell for a penny … Some of the boys who pursue this traffic are masters of all the trades that appertain to begging.

I have an instinctive desire to rescue the children from Mayhew's suspicion and scorn; for no doubt the flower sellers of Dhaka, too, would probably incur his resentment. Quite wrongly.

On the east side of the Parliament complex in Dhaka, just below a busy traffic intersection, Reena, who owns the flower stall, has tied a length of blue polythene from two trees to a couple of stakes on the edge of the sidewalk. The rain taps rhythmically on the material as Reena and her child helpers go about their work, making garlands, bouquets, floral displays in wet foam set in little wooden baskets.

Reena came to Dhaka from the north of the country when her father died in 1983. She worked as a maid in Dhaka. The lord of the house (as she calls him) was not good, and she was 'mentally tortured'. She left and sold peanuts on the streets. She sent for her infant son: she had been married at twelve. Reena is now thirty, her son seventeen.

Reena has been selling flowers for ten years on this spot. She is part of the landscape; part also of the pattern of intimidation and extortion by police and *mastaans* (thugs). Although she makes 2,000 taka daily, she pays the police 200. She also pays about a dozen boys who help her. The children do the actual selling at the traffic lights. Most come in mid-afternoon when the flowers have been prepared, and they stay until eleven or twelve at night. The flowers are mostly scentless crimson roses, *kathalthapa*, and sweet-smelling *rajni sughanda*, cream-coloured bells on long stalks.

Alamin is twelve. He has worked for about three years. Some days he makes 50 taka, others 70. Much depends on the weather, the day of the week and whether the police decide to harass them or not. Friday and Saturday nights are best. Alamin has three sisters. One is in a garment factory, another is a housemaid. His father is a beggar. Alamin has never been to school. Sometimes *mastaans* come and take money from the boys as they wait at the traffic lights.

The work is dangerous. A long uneven queue builds up at the traffic signals, and motorbikes roar menacingly between the stationary vehicles, disappearing in a streak of pollution and eddies of dust. The children learn to skip out of the way just in time. They tap on the windows of cars and vans, sometimes mime hunger at the occupants, touch the knees of passengers in baby-taxis and proffer bunches of sweet-smelling flowers. People occasionally seize the flowers and drive off without paying. Accidents also happen.

The boys are often arrested by the police. They are booked under the pretext of infringing traffic regulations, such as crossing the road when the lights are green – the most trivial excuses while Dhaka seethes with crime and political violence, monstrous violations of the law, which all go unpunished. Alamin has been arrested many times. There will be a fine of 200 taka, which effectively wipes out three days' work. He is not scared – it has become routine. Alamin gives all the money he earns to his parents. Before coming to work he plays cricket. Life, he says, is good.

Shopun is fourteen and has been a flower seller for three years. He came to Dhaka from Barisal after his father was drowned in the floods of the early 1990s. His father had gone home on vacation from his job as messenger in the Parliament building. Their small piece of land was eaten by the river, and everything was ruined. After his death, Shopun's mother took over her husband's job. She earns 1,500 taka a month. Shopun makes about 60 a day. He gives her half of what he earns. With the rest, he buys food, goes to the cinema with friends. Shopun says, 'This is not a good life. Education is important. I want to do something in the future.' He went to school, but left when his father died. Shopun doesn't like Dhaka and would prefer to be in the village.

Nobi is fifteen and comes from Mymensingh; he has worked for three or four years. His father is dead and his mother a maidservant. She goes to several houses each day where she cleans vessels, washes the floors and cuts vegetables. For this she is paid 600 to 800 taka monthly. Nobi earns 50 or 60 taka a day. He has two sisters, and they live in a slum area where they pay 300 taka a month in rent. He likes Dhaka because there is work here. He went to school, but left when his father died. When he is older he wants a regular job, on the railways. The first time he was arrested and taken to the police station he was very frightened, but now it is normal. In the mornings, he works at another flower-stand.

Shafiul who is twelve is Reena's nephew, the son of her brother. He has a brother and two sisters at home, but the family is landless. He lives with Reena and has never been to school. He earns between 30 and 50 taka daily, but takes nothing for himself since he owes everything to his aunt. He came here when he was eight. His ambition is to become the driver of a private car, like those that stop at the lights with their rich passengers in the back. Shafiul gets up at seven in the morning and goes with Reena to buy the flowers in the market. He helps her to carry them here and works all day.

Anwar is now sixteen. He comes from Mymensingh; his family are landless and his father too old to work. He is the only boy and has three sisters. They are all married. His mother is a maidservant, earning 200 to 300 taka a month (US$ 4–6). Anwar makes about 40 or 50 taka daily. He studied until class three. The family manages to eat three times a day, but when the police confiscate money his family goes hungry. Anwar gives his earnings to his mother. They live in a *bustee*, paying 200 taka a month.

In this conspicuous place the children are sometimes joined by other children passing by, to whom this work looks enviable and secure compared with the precarious livings some of them make. Rashida, eleven, is selling Cadbury's chocolate eclairs from a plastic jar. They didn't taste like Cadbury's – maybe they were counterfeit. Rashida goes to school from eight until two, then sells sweets until dark. Her mother is in a garment factory, her father dead. A thin child, without shoes, a grubby floral dress, she walks off through the evenly spaced trees in the park, her little legs spidery in the short dress.

Rupa sells water. She doesn't know her age, but looks about nine. She is the sixth of seven children. Her father is dead and her mother a domestic worker. She carries on her head a large white plastic container and a metal beaker, and sells at 1 taka a time. In the heat of may she makes 20 or 30 taka a day. She is small, with short hair, a dark ragged dress, without shoes. She gets the water free from a tubewell. She has two sisters working in garment factories and one brother who is a Tempo helper.

It is surprising how many children say their father is dead. Perhaps it is less shaming to say that he has died than that he has abandoned the family: in this sense, dead has a number of meanings. It may be they have been told he has died. It may be that he fails to keep them, has another wife, but they will not say so to strangers. No doubt, Mayhew would have found in this evidence of their mendacity. However, the low life expectancy in Bangladesh means that certainly a number of fathers will have died since mortality is high, particularly among cycle-rickshaw drivers.

By ten thirty in the evening the traffic has slackened. The boys still tap wearily on the windows of cars, sheaves of unsold flowers in their arms, faces lined with fatigue. Grimy, smelling of exhaust smoke, eyes red from the fumes, they droop poignantly like the wilting flowers they can no longer sell.

Chapter Twenty

Therese Blanchet says that 'the meaning of work and child labour cannot be grasped through economic criteria alone'. In other words, there are constructions of childhood which differ from Western norms; and these should be taken neither as an excuse to consent to violence and exploitation of children in the name of cultural pluralism nor as something to be extirpated at all costs in a missionary enthusiasm for the universalising of Western values. The balance between these is extremely hard to maintain.

Judith Ennew, in the *Handbook of Children's Rights* (1995), criticises the UN Convention on the Rights of the Child, observing that 'in the drafting process, the resulting text and in its implementation, takes as its starting point Western, modern childhood, which has been "globalised", first through colonialism and then through the imperialism of international aid'. This rather schematic view, of course, takes little account of what has been an even more profound influence upon the children of the earth – the vast spread of publicity of transnational companies, the cargo cults around certain consumer fashion articles; even the availability of the smallest luxuries in the remotest settlements on earth – from the appeal of Coca-Cola, Cadbury's chocolate, biscuits and soap in the most remote rural settlements, to Barbie dolls, computer games, the products of Disney and music cassettes in the cities.

Indeed, even at the heart of the contemporary Western world there is a fusion of old and new forms of exploitation and enslavement, and it may be discerned in everyday scenes in the cities of the West. In the ubiquitous shopping malls young people, starvelings of a global market, haunt the palaces of merchandise; girls and boys – black boys and girls who are the descendants of slaves or the escapees of Empire, white youth who are the great-grandchildren of those whose energies were devoured by mills and factories – covet the fashions and apparel, the shoes and accessories which bear the mark or logo of this or that transnational company. They are, in their way, also bonded, since their longing for all the desirable commodities is linked indissolubly to the involuntary enslavement of the children, their peers, young women and men in the garment and shoe factories of Dhaka, Jakarta or Ho Chi Minh City. The destiny of the children of privilege is joined with that of the helpless producers of their daily necessities, mediated by powerful entities who bestride the globe with their insolent triumphalism and whose imperial reach controls budgets greater than the GDP of Bangladesh. But that is another story – although it is also the same story.

In the most recent discussions of the subject, the outright abolition of the work of children is seen as an impossible objective, for both practical and humane reasons. The more modest aim now is that only the most hazardous and damaging labour should be banned.

This is perhaps symptomatic of the now almost unanimous belief in a benign gradualism, inscribed in the integration of all societies into the global economy. This, it is assumed, will, with time, more or less automatically float all countries out of poverty, provide all children with education and all adults with an income sufficient for them to dispense with the labour of the very young. In a choiceless globalisation, which is held to be as beneficent as it is inevitable, what other response is possible than a long-term laissez-faire which joins prudence with inevitability?

But for as long as the countries of the South must offer the labour of their people in the Dutch auction of global competitiveness, who is to ensure that those whose work commands the lowest price will be protected from hunger and want?

The fateful meeting of the World Trade Organisation in Seattle at the end of 1999 collapsed because the WTO was promoting itself as regulator, not only of world trade, but also of 'core labour standards' (as well as of environmental protection). It intended to ensure that no country should take advantage of such ancient economic devices as child labour, prison labour, bonded labour and all the other ingenious variants of slavery which persist in the modern world. Governments of the Third World fiercely resisted any such move since they saw in it a sinister form of Western protectionism – a determination by the powerful to prevent them from availing themselves of those spectacular abuses of humanity by means of which the West grew rich. No wonder the negotiations broke up in disorder and acrimony.

How to promote, on the one hand a worldwide system of 'deregulation', liberalisation and minimal government, whereby parliaments and legislatures cease to superintend the welfare, health, nutrition and education of their peoples (while still retaining responsibility for these things), and at the same time insist that they eliminate 'unacceptable' forms of labour, including that of children? Who is going to supervise such a prohibition, given the reduction in government functions now deemed indispensable for the creation of wealth?

In any case, as we have seen, most children are employed on their own account, by relatives, their own parents or small employers – to abolish child labour in farms, fields and workshops would require a quite impracticable intervention in the lives of families, which no one is seriously contemplating.

This is why, in the end, observers, analysts and humanitarians fall back upon pious and comforting formulae about 'education' as the supreme antidote to child labour. This, in the end, turns out to be yet another means of postponing the day of reckoning, as the millions of educated but workless young women and men in the Third World testify. They have dutifully passed through twelve or fifteen years of education, only to wind up as unemployed graduates, ambitious and qualified, but unfit for work, incapable of undertaking any productive occupation, unable to integrate themselves, no matter how 'integrated' their economy may be into the global market. What education? Education for what?, they ask – what purpose has been served by the rote learning, the superficial literacy, the orthodoxies of business culture, the degrees in marketing and management, the hi-tech know-how, the assembly-line acquisition of skills so flexible that they can be inserted into any demeaning low-paid function in the service sector?

In the meantime, many children are acquiring actual technical competence – sometimes brutally and dangerously – in the factories, workshops and garages of Third World cities. Who is to say that such training is worse than that of the aimless graduates who are often swept up in the service of criminal gangs, drug runners and crooked political businessmen, who recognise the value to their version of enterprise of youthful energies which the global market does not?

So what possible coherent response can be expected to such a confusing and comfortless situation? A shrug and a hope for the best. Dependency upon gradualism. More campaigns. More liberalisation. More controls. But in the end, faith in a future in which everyone will grow rich and child labour will wither away, as many of the actual children who perform it have already done.

The 'debate' has been put on ice, with the abolitionists promoting their vision of a labour-free childhood and the gradualists holding seminars in the luxury of five-star hotels all over the globe, while the international financial institutions showily advertise their recent conversion to poverty abatement and good governance; and the poor get poorer, and more children must enter the labour market for the sake of survival.

There is no satisfactory answer within the existing global arrangements. The global market no more caters now to the need for livelihood, education, food or health than it did a hundred years ago when William Morris saw a humanity under the banner of the

'world market', wasted either by poverty or excess. We look for
human well-being in the wrong place; surely, after all this time,
our puzzlement at not finding it there is disingenuous.

The further spread and intensification of the global market
economy is the only answer to wrongs and evils which that same
market has promoted. Who is going to explain it to the hungry,
wanting children of the world?

Mirpur 11 in Dhaka: a place of slums, industrial barracks and
workplaces. Narrower roads and even narrower alleys, where the
houses are so close together that people can scarcely pass between
them. The most imposing buildings in the neighbourhood are the
garment factories, seven- or eight-storey brick towers rising above
the huts, their white strip-lighting blazing out into the dark, palaces
of production over the hovels of survival.

Here is a community of former Hindu Untouchables, shoe and
leather workers, known in Bangladesh as *muchis*. After
Independence, they were rehabilitated and housed in plain huts
of bamboo, with woven bamboo walls and earth floors. Many of
these houses are now dilapidated. The people are among the
poorest in Dhaka. Men and boys make a pitiful living at the
roadside, in the market, at intersections in the slums, polishing,
mending, making, repairing shoes and chappals. Women are more
or less confined to the house.

I wanted to meet the leather workers because of my own family's
connection with the same industry. I don't know what points of
contact I expected to find – certainly the conditions of living could
at first sight scarcely have been more different – the rented red-
brick houses in a mossy cold street in Northampton bear no
resemblance to the dusty *chetai* houses and the mauve water
hyacinths in the stagnant pools around them.

But the parallels exist in the expedients and strategies of survival
of the poor, the patterns of thrift, the use of native intelligence to
outwit the guile of the powerful, the hatred of moneylenders and
landlords, the strength of women, who always remain with the
children while men so often run away in despair. The *relationships*
of oppression are the same; the face of suffering is recognisable
whatever its ethnicity; hunger assumes no exotic guise, material
scarcity does not dissimulate itself behind unfamiliar cultural
practices; endurance and stoicism are known in the same way; and
the lineaments of hope do not vary from one country or one time
to another.

The faces of the young women I met in Mirpur 11 evoked the
sepia and fading prints of the seven girls my mother and her sisters
had been, as I had never known them, but as they remained,
battered and dog-eared in the little shoe-box of remembrance to
which all our small family treasures had remained – May, unaware

of the leukaemia that was to claim her; Em knowing nothing of the TB that would take her husband; Glad, oblivious of her husband's future syphilis; all smiling in the sunshine, while the shadows of the houses behind had not yet reached the patch of light where they stood ... Their youth, too, had been used up by work, their intelligence suppressed by the demands of survival, their spirit subdued by the indifference and prejudice of those who still thought of themselves as their betters.

In Mirpur, I went to a small hut, a class run jointly by the Association for the Realisation of Basic Needs and the International Labour Organisation in Dhaka. This modest effort was designed to give girls a skill to enhance family income – sewing and stitching, a safe, domestic occupation.

Meena is sixteen, self-employed and married. She does zari work, intricate embroidery for the borders of sarees in gold, silver and scarlet thread. Her long hours spent on the traditional motifs yield 60 taka for a dozen pieces (US\$ 1.20) – three days' work. Meena could have earned more in a garment factory, but girls from her community will not go for such jobs because they fear rejection, not only by the manufacturers but also by the other workers. Meena's three brothers mend shoes on the pavement. She was lucky – she was taught zari work when she was ten and has been earning a kind of living for five years.

Deepu does the same work but payment is irregular. The employer cheats her, so that she receives 400 to 500 taka a month (about US\$ 10) for work that ought to bring twice as much. She has four brothers, but her father is too old to labour. An older brother is 'defective', her mother an invalid. Deepu takes responsibility for all the household work. The family left their native place after Independence, having neither land nor homestead. Her ambition is to have a machine at home for tailoring. A garment factory is not 'safe'; this means she would be sexually at risk from male employees.

Sagori, fifteen, is one of four sisters and three brothers. Her father died of cirrhosis of the liver and her mother is almost blind. Her brother is a shoe cleaner who knows the most profitable spots to work. He is twelve, and earns 20 to 30 taka daily (about 50 US cents). There are nine people in the family, whose three working members are all children. Their total earnings are less than 100 taka a day (US\$ 2). Sagori is responsible for fetching water – the supply line is distant, and water comes only for two hours in the afternoon. Sagori prepares food, cleans the house and looks after the younger children. She gets up at six o'clock and, if there is food in the house, prepares breakfast. Today there was nothing. If the brothers have earned money by noon, 'there will be cooking'; if not, they will wait until evening. Sometimes, their meal is rice, chilli, salt and plain water. In the lean season, they borrow from moneylenders, at a rate of interest of 10 per cent per month.

Taramani is twelve, the eldest of four sisters and two brothers. Her father repairs shoes on the footpath and brings home 40 to 60 taka a day. No one else in the family is working. This morning they ate nothing. The walls of the house are broken, the *chetai* is bleached and brittle, but they cannot afford the repairs. Taramani does not know if there will be food at midday. She feels hunger most days, and when the children cry she is ashamed because she must comfort them with false promises such as, 'Hush, Daddy will come and he will bring food.' Taramani is desperate to learn a skill. 'I'll do anything, but I cannot go into a factory, or the people will say I have gone astray, I am meeting boys, and then my chance of marriage will be spoiled.'

None of the girls has been to school for fear of being molested. The people of the community are so despised that miscreants feel free to do with them whatever they choose. The best hope of a girl is to be 'given' in marriage to a boy earning more than 40 taka a day.

Zamiti is twelve, the fifth of seven sisters. Her parents are blind. Her oldest sister was married, but has been abandoned by her husband. She has now gone to be a sweeper in a factory, earning 350 taka a month (US$ 7). Another sister has become a maidservant. She receives no salary, only food and clothing. This is a bold step. Most girls may not even work as maidservants for the same reason they do not go into the factories. They fear they may be abused or raped. Zamiti's sister is ten. She comes home late at night, and her family are waiting for the leftover food which is her wage.

The people are settled along dusty paved walkways beside a deep pond, in which the early spring water is low, and has been set with rice plants. The *chetai* walls, exposed to too much sun and too many monsoons, need renewing, the metal roofs are flaking with rust. The interiors are bare, devoid of possessions: a dusty grey emptiness so complete, it is difficult to believe that ten or fifteen people sleep in each one – some utensils, dingy clothes on a string, a piece of cloth that serves as a baby's sling, a bedroll. On the threshold of each hut, a grey clay stove, blackened by fire smoke. Some small children in the lane are playing with bottle tops, a threadbare broom, a chappal with broken thongs. New meanings of poverty must be formulated for these desolate urban places. Existing words cannot convey the epic misery and the gaunt raw beauty of these young girls, excluded even from the most despised occupations of other children because of an antique prejudice and the ghost of a caste system in theory long abolished.

Most of the families are large. The middle class complain of 'population', but it is easy to see how readily the families of the poor are depleted, used up by overwork, malnutrition and sickness.

The renewal of flesh and blood, and its capacity for work, is the only defence against utter destitution. Those who complain of population – and who does not? – have little understanding of the fragility of the lives of the poor. Yet we knew this not so long ago. My mother was the youngest of thirteen children. We know that social security is the surest way of bringing down the size of families – a guarantee that their young will survive. But social security has been written out of the drama of globalisation as interference with the market.

I was reminded of a story I had heard from an old miner in Wigan some years ago. He had a large family, ten or eleven children. One day, he heard a neighbour complain that people should not have so many children. He called his children into the yard, lined them up in a row. 'Now,' he challenged the neighbour, 'look at them, and tell me which ones I shouldn't have had.'

We in the West have forgotten so much under the influence of our slender and threatened privilege; even though it is only yesterday that our parents and grandparents told us stories which are familiar to the children here in Mirpur – sorting out the rotting fruit in the market late in the evening for a few mildewed oranges or overripe bananas; waiting in the early morning at the bakery door for the previous day's stale pastries; not eating until the parents came home at night with a penny bloater; not knowing if there would be a day's work tomorrow; standing in the factory yard or at the dockside, waiting like cattle for a foreman to appraise your muscles; children taken out of school on their twelfth birthday and starting a factory job the next day, little girls tying knots, mending threads, sweeping up the waste under the looms and machines; children scavenging for rags, bones, rabbit skins or old iron in a hessian sack slung over the shoulder; taking starchy meals in a basin to brothers and fathers in their noon factory break; picking a few bluebells, mushrooms, blackberries or hazelnuts from woods and meadows for a few pennies. No wonder the ties of kinship have been loosened in our country; if they had remained, we might have recognised our kindred elsewhere, in these children, in the sombre, forsaken places of the world.

One last, defiant emblem of the work of children in Bangladesh. In north Bengal, close to Bogra and the ancient Maurya ruins of Mahasthanghar, two boys are working. They are repairing embankments around the paddy fields in readiness for the *aman* (rain-fed) crop. It is a sunshiny day, with a breeze through the mango and jackfruit trees. In some fields, new rice has already been transplanted, in others, it stands in close-sown emerald patches, waiting to be thinned. Nizam and Moznu are both sixteen. Nizam left school at twelve. His family owns 14–15 bighas of land. Moznu left school at nine. He didn't like study and wanted to work in the fields. His family also have 15 bighas of land.

The two boys are barefoot, their legs covered in mud which has dried to pale grey at the knees, still wet around the feet. Their hands are gloved with the same dry earth, and mud has splashed its exclamations on their faces. Each wears a blue *lunghi* tied up at the waist to protect it from the water. Both are sturdy, with big veined hands and an arrow of dark hair at their chest. Their muddy faces are gashed with a silver smile. They exude an unmistakable well-being. They spend their free time in *adda*, which means gossip, cinema and TV, meeting their friends at the village tea stall, roaming in the fields. Why did their parents not encourage them to stay at school? The smiles widen in pity for the ignorance of the question. It is quite clear that the parents do not need the labour of their children – they are well-to-do farmers. Work is simply what they do. Why look further for an occupation than these rich paddy fields? The boys are secure. *Dhan* means paddy and also wealth. They know how to do everything necessary to keep their land productive. Their future – work, marriage, children, land – stretches ahead of them like the straight open plains of north Bengal with its perpetual harvests. What can education teach them that they do not already know? They do not reflect on their life, but merely live it; feet in the earth of Bengal, growing heedlessly in the fields as the wild flowers with which they share the landscape.

Bibliography

Aries, Philippe, *Centuries of Childhood*, Random House, 1965

Blanchet, Therese, *Lost Innocence, Stolen Childhoods*, The University Press, Dhaka, 1996

Bloom, David and Sachs, Jeffrey, *Geography, Demography and Economic Growth in Africa*, Brookings Papers on Economic Activity, no. 2, 1998, Washington DC

Briggs, Asa, *The Age of Improvement*, Longmans, 1959

Engels, Friedrich, *The Condition of the Working Class in England in 1844*, George Allen and Unwin, reprinted 1952

Ennew, Judith, in Franklin, Bob (ed.), *Handbook of Children's Rights*, Routledge, 1995

Gaskell, P., *The Manufacturing Population of England*, 1833, Ayer Co. edition, 1972

Gatrell, V.A.C., *The Hanging Tree*, Oxford University Press, 1994

George, M. Dorothy, *London Life in the Eighteenth Century*, Kegan Paul, Trench, Truber and Co., 1925

Green, Duncan, *Hidden Lives*, Radda Barnen, 1998

Gregg, Pauline, *A Social and Economic History of Britain, 1760–1980*, Thomas Nelson & Sons, 1990

Hammond, J.L. and Hammond, B., *The Rise of Modern Industry*, Methuen, 1947

Harrison, J.F.C., *The Common People*, Fontana, 1984

Horn, Pamela, *Labouring Life in the Victorian Countryside*, Gill and Macmillan, 1976

Howitt, William, *The Rural Life of England*, Longman, Orme, Brown, Green and Longman, 1838

ILO Survey, Geneva, 1996

Life of Sir Samuel Romilly, John Murray, 1842

Mayhew, Henry, *London Labour and the London Poor*, Dover Publications, 1983

Pendergrast, Mark, *For God, Country and Coca Cola*, Weidenfeld and Nicholson, 1993

Pereira, Winin, *Asking the Earth*, The Other India Press, Goa, 1992

Quennell, Peter (ed.), *Mayhew's London Labour and the London Poor*, Bracken Books, 1984

Rohfitsch, *A Critical View of the Judicial Institution in Relation to the Rights of the Child*, Radda Barnen, 1995

Rowntree, Seebohm and Kendall, May, *How the Labourer Lives*, Thomas Nelson & Sons, 1913

Second Report of Children's Employment Commission, 1843

Skrobanek, Siriporn, *The Traffic in Women*, Zed Books, 1997

Thompson, E.P., *The Making of the English Working Class*, Victor Gollancz, 1963

Unwin, Mrs Cobden (ed.), *The Hungry Forties*, T. Fisher Unwin, 1904

Walvin, James, *Black Ivory*, Fontana Press, 1993

Williams, Eric, *Slavery and Capitalism*, North Carolina University Press, 1945

Index

Compiled by Sue Carlton

abolition 14
  gradualist approach 40,
    67–8, 150, 151
  and protectionism 22, 48,
    54, 150
  of slave trade 2, 3, 14
  see also child labour,
    reform/abolition
Abu Dhabi 138
adulthood, effects of child
  labour 112–16
Africa, migrant workers 9
agricultural revolution 56–7
agricultural work
  child labour 23–4, 56, 71–3,
    74–5, 77
  hazards 73
Alam, Shahid-ul 118, 119,
  120
Amazon 67
America, cotton mills 15–16
amphetamines 8
apprentices 4, 15, 122–3
  unpaid 42–3, 69–70, 76,
    122
Aries, Philippe 80–1
Association for the Realisation
  of Basic Needs 153
Awami League government
  35, 60

banana selling 119
Bangabazar 107
Bangladesh
  activists 57
  begging 94–5, 133–4, 135,
    143

child labour 5, 28–9, 64,
  121–8
and conception of childhood
  80–1
disabled people 31
domestic servants 98–103
economic expansion 67
economic pressures on
  parents 7
education 30–7, 39–40, 52,
  112, 125
factory working conditions
  16, 27–8, 33–4, 76–7, 125
  see also hazardous occupa-
    tions
garment factories 7, 16,
  27–8, 76–7, 142, 143
and global economy 32, 38,
  49
and Harkin Bill 21–2
health 34–5, 36, 45, 60
hospitals 36
housing 35, 36, 124, 154
and inequality 28, 31–2, 33,
  34
life expectancy 148
living conditions 35, 36, 61,
  124
migration 10–11, 24–5, 44,
  84, 85, 112–13
orphans 55, 94–6, 108
poverty 34–5, 36
and protection of children 18
street children 31, 55, 94–5,
  107, 129
treatment of poor 25–6
wealth creation 19

Bangladeshi National Party 35
Barbados 14
Barisal 77, 84–9
  migration from 27, 60, 69,
    141, 146
Barnardo, Dr 130
begging 94–5, 132–3, 135,
    143, 145
  and amputation 133
  attitude of rich to 133–4
Begum, Jehanara 21
Bell, John 81
Bengal 67, 155
Beribund 58–9, 60
Beribund School 32–3
Bhuiyan, Abdur Rauf 38–40
bidi factories 23, 65
biscuit making 85–6
Blanchet, Therese
  concept of childhood 80,
    81, 82, 149
  domestic servants 22, 58,
    99, 101, 102–3
  and reform 23, 65
  street children 04–5
Boards of Guardians 7
Bogra 73, 143, 155
bone-pickers 91
Booth, Charles 108–9
Brazil
  child labour 5, 53
  expansion 66–7
  murder of street children
    54–5, 130
  urbanisation 54
brick-breaking 34–5, 36,
    58–61, 69, 88
brickworks 61–2
Briggs, Asa 71
British Guyana 17–18
Buriganga river 127–8
Burma 50

Camel Jockey Association 139
camel jockeys 138–9
Candler, Asa 5

caste system 152–4
Chakmas 52, 67
Chamorro, Violet 49
Chateaubriand, René 104
child brides 114–15
child criminals 81
child labour
  19th Century London 90–1,
    96
  age limits 22–3
  agricultural work 23–4,
    71–3, 74–5, 77
  in Barisal 86–9
  and child work 64–5, 67–8
  consulting children 65, 68
  and cultural difference 68,
    80–3, 96
  defence of 1, 4, 5, 9, 18, 48,
    64
  and discipline 27–8
  exploitation 38, 39, 65
  factory work *v.* agricultural
    work 56
  and health 6–7, 28, 40, 44,
    60
    see also hazardous occupa-
      tions
  in home 81
  Income Tax Office 97
  infants 7, 36, 59
  and informal education 30,
    39–40
  investigations into 64
  keeping awake 7–8, 16
  learning skills 38, 40, 48,
    68–9, 112, 124, 151
    see also apprentices
  official indifference to 29,
    64
  positive elements 38, 39, 48,
    64–5, 71, 118
  and poverty 5, 7, 10, 48, 51,
    69
  reasons for 50–1
  reform/abolition 5–6, 28–9,
    47–8, 64, 150–1

consequences of 21–3,
38–9, 53
and education 31
gradualist approach 40,
67–8, 150, 151
and increasing wealth 19
regulation 69, 150
and size of child population
71
and western media 53
and Western norms 39, 149
workers becoming adults
112–16
working hours 4, 17, 48, 64
see also domestic servants;
Kamalapur Railway
Station; working
conditions
child work 64–5, 67–8
childhood
conceptions of 39, 80–1,
102–3
globalisation of 149
children
attitudes to work 65, 68,
81–2
as individuals 89
inventiveness of 107
relationship with parents
88–9
and religion 96
removal from families 9, 44,
57, 73, 83, 122
rights of 29, 48, 50, 51–2,
80–1, 149
trafficking of 12–13, 50,
138–9
see also camel jockeys;
sexual abuse
Children's Employment
Commission (1843) 6,
122
chimney sweeps 137, 139
Chittagong 11, 132
Chittagong Hill Tracts 52, 67
Christianity 96

coal mines 7, 8
Cobbett, William 56
Coca-Cola 5
colonialism 20, 66–7
consumerism 50–1, 149–50
convicts, and labour 14
corporal punishment 16, 75,
76
cotton mills 1, 4, 6, 7
America 15–16
cripples 134
crossing-sweepers 131

Davies, Rev. David 71
debt 49, 50
Defoe, Daniel 4
deprivation, strengthening
effect of 117
Dhaka 11–12
agricultural work 75
brick-breaking 58, 69
child labour 121–8
domestic servants 98–103
garment factories 7, 16,
27–8
International Labour
Organisation 153
Kamalapur Railway Station
129, 130–1, 132, 134–7
police 96, 130, 133, 135,
146
slums 109, 111, 140, 152
Sutrapur area 96–7
vehicle repair shops 40, 41,
42–3, 69, 70
Dhaka Electricity Supply
Undertaking (DESU) 40
discipline 26–8
divorce 115
Doloikal 40, 41
domestic servants 73–4, 82–3,
98–104, 144
abuse of 22, 99, 154
and alien values 100–1
changed names 101
and education 100

domestic servants *continued*
  exploitation 99, 102, 123
  importance of age 102–3
  and sense of identity 100–1
  untrustworthiness 103
  working hours 99–100
DRIK 117, 118–19, 120
drugs 105
Dubai 138

East India Company 19
East Timor 66
economic growth 19–20, 67
education 30–7, 39–40, 52,
    143, 147
  antidote to child labour 31,
    151
  benefits to employers 44
  domestic servants 100
  and employment 38, 112,
    125, 151, 153, 156
  and empowerment 121
  evaluation of 52, 65
  flexibility 35, 39–40, 130
  importance of 60–1
  informal 30, 39–40
  and parent-child relation-
    ship 120
  right to 51
  selection 31–2
  skills training 153
  street children 130, 132
  see also Underprivileged
    Children's Education
    Programme (UCEP)
egg selling 119
electricity supply
  failure of 40, 42
  illegal 109
employers
  and child workers education
    44
  educating 39
  providing skills training 38,
    40, 68–9, 112
  see also apprentices

empowerment 65–6, 121
Engels, F. 61, 105, 111
Ennew, Judith 149
Equiano, Olaudah 13–14
Ershad, General 67
Essex 75

Factories' Inquiries
    Commission (1836) 6
Factory Acts 5, 17
Factory Commission (1833)
    64
factory work, symbol of
    modern world 57
Faridpur 70, 75, 113, 128,
    136
fertilisers and pesticides 78,
    113
Fielden, John 15–16
financial institutions 49
  and tolerance of child work
    68
flower sellers 145–7
fodder, collecting 74, 75
forest, destruction of 52, 67
free labour 2, 3, 79

garment factories 7, 11, 142,
    143
  and Harkin Bill 21
  working conditions 16,
    27–8, 76–7
Gaskell, P. 57
Gatrell, V.A.C. 81
George, Dorothy 70
glass factory, Bangladesh 33–4
globalisation 18–19, 47, 49,
    67, 150
  consumerism 50, 149–50
  and economic adjustment 49
gradualism 40, 67–8, 150, 151
Green, Duncan 48, 53, 129
Green Revolution 84
Gulshan, Dhaka 73

Hamden, Mr 18

Hammond, J.L. and B. 3–4, 8–9, 56–7
handloom weavers 7
Harkin Bill 21, 22, 39, 53, 64
Hart, Richard 13–14
hartal 11
Hasina, Sheikh 35
Hawkins, Dr 6
Hayek, Friedrich von 19
hazardous occupations 40, 41–2, 44–5, 98, 121, 136
    see also brick-breaking; camel jockeys; NAYAN; working conditions
health, medical expenses 45
Honley 76
Howitt, William 71–2
humanitarianism
    and increasing wealth 19–20, 23, 47, 139
    see also wealth creation
Hungry Forties 72

illiteracy 52
    and innocence 110
ILO (International Labour Organisation) 68, 121, 153
    Convention on Child Labour 64, 67
Income Tax Office, employing child labour 97
India 5, 52, 112
Indonesia 66
Industrial Revolution
    and child labour 1, 2, 3, 55, 71
    and free labour 2, 3
    and life of poor 8–9, 25–6
    and social injustice 25
    and urbanisation 54
inequality
    Bangladesh 28, 31–2, 33, 34
    in global economy 68, 150, 151–2

International Monetary Fund (IMF) 49
Irian Jaya 66
Islam 96
Islambagh 121–7

Jamaica 16–17
Jessore, trafficking children 12

Kamalapur 69
Kamalapur Railway Station 129, 130–1, 132, 134–7
Kendall, May 23
Khulna 12, 72–3, 112–13
kidnapping 12, 14
    see also children, trafficking of
Kishorganj 132
knot-tiers 82
Koylaghat 127
Kustia 23

Lancashire 7, 111
Latin America, street children 54–5, 129–30
leather workers 152
Leicestershire 73
Liberation War, Martyrs' Monument 35, 36, 60
Litton Mill, Derbyshire 14
London
    Saturday night markets 106–7
    slums 108–9
    workhouses 1, 3–4
    see also Mayhew, Henry
luxury goods, and child labour 149–50

Madaripur 120
Mahrashtra 52
Manchester 6
Maresworth, Hertfordshire 72
markets 106–7
Marks & Spencer, and child labour 53

Marx, Karl 22–3
Maryland 14
mastaans 146
match sellers 133
Mayhew, Henry 90–1, 96, 130
    beggars 132–3, 145
    cripples 134
    crossing-sweepers 131
    mudlarks 92–3, 128
    Saturday night markets
        106–7
    street vending and begging
        145
    unreliability of servants 103
media, exposing exploitation
    53
Memorandum of
    Understanding 64
metal, galvanising 44–5
metal industries, Midlands 6–7
migration 10–11, 33, 57–8, 84
    seasonal 61–2
    transmigration 66
    see also children, removal
        from families
minimum wage 1
Mirpur 100, 103, 117–18,
    140–1, 152–3
Mohammadpur 32–3, 35
moneylenders 154
moral crusades, consequences
    of 53–4
Morris, William 151
muchis (former Untouchables)
    152–4
mudlarks 92–3, 108, 128
Mumbai 12–13
Muslimbagh 128
Mymensingh 135, 141, 147

Narayanganj 91, 113
National Child Labour
    Committee 5
NAYAN foundation 38–40,
    44
    resistance to 45–6

Nepal 5
New Market, Dhaka 135–6
Newton, John 3
NGOs
    in Bangladesh 28–9, 30, 47
    and child labour 67
    loans from 36
    rescuing camel jockeys 138
    and street children 130
Nicaragua 49
Norfolk 75
Northamptonshire 77, 82, 152
Nova Gerusalemme 54–5

Opposition Bangladeshi
    National Party 11
orphans 55, 94–6, 108
Owen, Robert 7

Pahile Boishak 129
parents, economic pressures
    on 7, 50–1, 69
Pereira, Winin 52
Phensidyl 105
'piecers' 7
Pitt, William (the Younger) 1
plastic sandals factory 125,
    126–7
Plymouth 75
Poor Law Amendment Act
    (1834) 17
poverty
    alleviation 65–6
    consolations 105
    and cultural difference 54,
        80, 107–8
    and family size 154–5
prostitution, children 12–13,
    50, 114

Quennell, Peter 110–11

Rahman, Anisur 22
Rahman, Dr Wahid ur 67–8
Rahman, Helen 21, 99, 100,
    102

Rahman, Zia-ur 67
Rajshahi 45
Ramsey, James 26–7
Raybazar 60
Razzaque, Abdur 22
rickshaws, pushing 96–7
river erosion 120, 127, 147
Rohfitsch, 81
Romilly, Sir Samuel 9
Rowntree, Seebohm 23
rubbish recycling 87–8, 91–2,
    110–11, 125–6, 140–2,
    143

Sadaraghat River Launch
    Terminal 93–4, 95–6
Sadler, Michael 17
Safesave 61
Sandinistas 49
Satkhira 112
sexual abuse 114, 154
Sheraton Hotel 145
Shimu, Shima das 12
shoe factories 149
    Northampton 82, 152
SHOISHOB 21, 98, 99, 100
Siddiqui, Tanbir ul Islam 30, 31
Sierra Leone 2
Skrobanek, Siriporn 50
slave trade
    abolition of 2, 3, 14
    and children 2–3, 78, 104
    and cotton trade 18–19
    defence of 1, 9
    reform 17–18
    and white labour 14
    working conditions 16–17
slaves
    craftsmen 78–9
    and discipline 26–7
    and family life 78–9
    in modern world 139
    'seasoning'/breaking-in 63
    untrustworthiness of 103–4
slums 108–11, 113, 140, 152
    and self-reliance 110

Society for the Suppression of
    Mendicity 133
South America, migrant
    workers 9
Staffordshire 7
street children 31, 53, 54–5,
    94–5, 107, 129–37
    and education 130, 132
street vending 23, 145
structural adjustment
    programmes 49
sugar cane cutters 53
Sussex 74
Sutrapur, Dhaka 96–7
sweet selling 142–3, 147
Sylhet 135

Tamils 10
tanneries 125
Tempo taxis, helpers 135–7,
    142
Ten-Hour Bill (1847) 17
Thackrah, C. Turner 6
Thailand 50
Thompson, E.P. 7, 19, 21, 56,
    64, 81
thread-cutters 82
tokai 91
    see also rubbish recycling
transmigration 66
transnational companies 149
travel terminals 94, 108
Trinidad 17–18
truck drivers, helpers 45

UBINIG 12
Underprivileged Children's
    Education Programme
    (UCEP) 30–2, 36–7, 60
    and employers 31
    flexible school day 35
    selection process 31–2
    technical education 30, 37
    see also education
UNICEF 68

United Arab Emirates (UAE), camel jockeys 138–9
United Nations Convention on the Rights of the Child 29, 50, 51–2, 80–1, 149
United Nations Development Index 52
United Nations Special Rapporteur 139
Untouchables 152–4
urbanisation, and social disintegration 54

vehicles, reconditioned 40, 41, 42–3, 69, 70
Vikrampur 127

Walvin, James 2–3, 16–17, 28, 63, 78–9, 103–4
water hyacinths 88
water sellers 148
wealth creation 15, 47, 56, 66, 68, 139
  see also humanitarianism, and increasing wealth
welding 87
West
  consequences of interference by 53–4, 120

and wealth 66
young people 79
West Indies, slavery reform 17–18
Whitbread, Minimum Wage Bill (1796) 1
Wigan 17, 155
Williams, Eric 14–15
women
  from *muchi* community 152–4
  and Industrial Revolution 8
  trafficking of 50
working conditions
  19th Century Britain 6–7, 15, 17, 34, 35, 44–5
  Bangladesh factories 16, 27–8, 33–4, 76–7, 125
  slavery 16–17
  see also hazardous occupations
World Trade Organisation (WTO) 150

Yorkshire 7

zari work 153